The Universal Attraction of Gravity

By

Timothy Michaels

Table of Contents

Table of Contents 2

Preface.. 7

Chapter 1 - Introduction 9

Chapter 2 – The Beginning 11

Conservation 11

Uncertainty 13

Tunneling.. 14

Gravity 16

Force = Energy / Distance 16

Probable Distribution.. 17

Size of the Initial Universe 19

Minimum Number of Particles 19

Process of Distribution.. 21

Chapter 3 - Gravitational Interactions 23

Rotation.. 23

The Spiral.. 24

Force Of Gravity.. 26

Inverse Square Law 26

Masses 27

Creating The Formula... 27

Chapter 4 – Condensing Matter 31

Planets.. 31

Stars 32

Neutron Star 34

The Schwarzschild Radius 35

   Deriving the Formula 36

Chapter 5 - Black Holes 41

Outside 41

Inside 42

Chapter 6 – How Gravity Attracts.....................47

Communicating Gravity 48

Potential Energy 48

Radiant Energy 53

Force of Energy...54

LIGO 55

Gravity Is A Force...55

Wave Alignment 57

Alignment..59

Chapter 7 – Gravitational Dilation 63

Movement Theory..63

Dilation From Travel...66

Dilation From Gravity...68

Traveling vs Vibrating 70

Gravitational Heat 71

Chapter 8 – How Gravity Converts Matter Into Energy...........73

Particle Requirements 74

Formula for Gravitational Dilation......................75

How Gravity Waves Convert Matter Into Energy.....................77

Wavelength/Energy Relationship.........................78

Chapter 9 – Escape from the Inescapable 81

   Communicating Out of a Black Hole 82

   Matter Can Escape 83

Chapter 10 – Canceling Gravity 87

   More Than Acceleration 87

   Centrifugal Force Doesn't Cancel Gravity 88

   Gravity Cancels Gravity 89

   Hafele And Keating Experiment.................................. 91

   Temperature 93

   Gravity of the Universe.. 93

   How Gravity Cancels... 94

   Gravity Is Not An Object... 95

Chapter 11 - Orbits 97

   Orbiting a (Singularity) Black Hole............................ 97

      With Dilation 99

      Combining Dilations 101

      A Note About My Hafele and Keating Data 103

   Even Density Black Hole.. 104

Chapter 12 – Gravity is Soft... 109

Chapter 13 - All Orbits Accelerate 115

   Energy Shift.. 115

   Doppler Effect 117

   Migration 118

   Consequence... 119

   Earth... 120

Chapter 14 - Consequences of Acceleration............123

    Dark Matter 123

    Converting Matter into Energy 124

    Identity Crisis 124

    Dark Energy 125

Chapter 15 - Limit of Universe Expansion..............127

    Our Place in the Center 129

    Lifespan of the Universe 134

Chapter 16 - The Beginning and the End................137

    Microwave Background 137

    Space Growth 138

    X-ray and Neutrino Background 142

Chapter 17 - Conclusion 145

About the Author...147

Other Books by Timothy Michaels 149

Bibliography...151

Preface

After exploring absolute relativity, I wanted to look more closely at how it works with gravity. What I found was that it appears to solve some mysteries and also provides interesting predictions. For example, it can help us find where we're located in the universe and why it seems to expand.

While developing these ideas, I also found that there's more that can be explained about gravity using information we already have. We seem to know enough, now, to determine what it is and how it works.

Gravity is known as one of the four forces of nature. It's referred to as the weakest, and it's also the most obvious in our daily lives. It's said to exist as a particle like the other three forces: electromagnetic, strong nuclear, and weak nuclear. But it's the only one whose particle has never been found.

Some scientists believe that gravity is not a force at all, and is instead, a warping of space and time. But I believe my theories of movement and absolute relativity provide a more accurate picture of the world. They provide a model in which space and time are simple, fixed measuring systems, just as we experience them. As a force, I believe it has a very interesting and simple explanation. I'll be presenting how I believe gravity is an effect of energy and arises from a more fundamental law of nature.

My purpose in this book is to expand on the applications of absolute relativity by applying it to gravity, and also to introduce some new ideas about gravity. I will do this while presenting a cosmology, or history of the universe, based on these ideas to demonstrate their application.

I'll be applying some well-established physical theories as well as my own ideas, and will show how they all make sense together. And I'll provide reasons for all of the concepts which a reader may question, or which give interesting insight into gravity.

Some of the ideas referred to here are described in my previous books, *Out of This World: The Movement Dimension* and *Absolute Relativity: How Newton and Einstein Agree*. Other ideas are being presented for the first time. I will try to explain each concept completely enough so that they can be understood without the need to read the other two books.

Other concepts that are important to this book are quantum probability, probable distribution, and conservation of matter and energy. As we encounter each of those, I will provide some explanation of what they are so that a variety of readers can follow the concepts. There is sometimes a lot of math shown, but I believe any reader who chooses to skip those portions can still understand the ideas well enough to be able to judge their correctness.

Together, my findings indicate what the early universe may have been like and what its future may be. They may also solve some mysteries along the way.

I hope readers find this book interesting and enlightening, with its new ideas worthy of their consideration.

Chapter 1 - Introduction

I'm going to tell the story of the universe from its very first moment to its very last, in a way that I believe events have and will occur. Of course, I cannot cover every aspect of it. My focus will be on showing gravity's role in the whole process, because I believe it's much more significant than is popularly understood.

The reason gravity is a big factor in our universe as a whole, is that it is the only force which significantly influences the movement of bodies on the largest scale (the movement of galaxies and their stars).

The four known forces in our world seem to be the cause of all of our physical events:

- There are two nuclear forces (strong, weak) that have a very short reach. They are believed to have no influence beyond the size of an atomic nucleus. The strong nuclear force only extends for one ten-trillionth of a meter. And the weak force reaches one ten-thousand-trillionth of a meter.

- The electromagnetic force, which we experience as magnetism and electricity does have indefinite reach. It can have influence across the universe. But it's generally only locally effective. The reason for this is that it's made of two charges which cancel each other, and there's an equal amount of each in the world.

  The force of the north end of a magnet reaches across the entire universe. But so does the force of its south end. If there was uniformity in their arrangement this force could be significant. But every atom and every planet and star acts as a magnet. And they're not all oriented the same way. This results in magnetic forces often canceling each other. I believe this is the main reason why magnetism doesn't usually have influence very far. Magnetic forces aren't organized as uniformly throughout the universe as they are in a piece of lodestone (a magnetized rock).

- The fourth force is gravity. Gravity doesn't cancel in the same way as electromagnetic. There is no anti-gravity in the forward direction of time like there is an anti-north (or south) magnetic pole or an anti-positive (or negative) electric charge. All things are attracted to all other things by gravity. So I'll be showing how this force moves our world on the largest of scales.

We'll begin with the initiation of the Big Bang, even before the first moment of existence. You might say we're starting before the beginning. We'll see how the bang of the Big Bang occurred by the same fundamental principle which I believe causes the effect of gravity. Then we'll focus on how the gravity we're familiar with has directed our universe so far and will see it through to its conclusion.

Chapter 2 – The Beginning

If the universe began as a singularity (an object of zero size) it could never have existed as one. I define existence as being present in space and time, because this is where our laws apply. This early in the universe we can expect to encounter unusual concerns like this. But in fact, existence is an important consideration in solving many problems. We'll see, later, that being in or out of space and time are critical factors in the way the world works.

Since we see evidence of the universe expanding, it's easy to conclude that it was smaller in the past. Some of this evidence is in the form of the Doppler effect on the colors of stars. Just as the pitch of a passing siren changes, the color of light does too when it comes from a star which is moving away from us. This leads to the question of how small it may have been. So, we consider that maybe it was infinitely small, a singularity.

Since a singularity is an object of zero size, this means it must have infinite density. And according to Heisenberg's uncertainty principle, if it has zero size it must have infinite energy (or mass). This is a tricky concept, but we'll see how this works shortly.

In this state of no size and infinite density nothing can occur. The universe cannot expand. Contrary to our world experience, the more energy or mass something has, the smaller it is. The uncertainty principle tells us that an infinite amount of energy is perfectly content in no space, and that it in fact must be of no size.

I don't believe there's any way to make a universe out of this. In fact, this doesn't even describe our universe. It couldn't possibly exist as a singularity because that would violate one of the laws of nature.

Conservation

One of our most fundamental physical laws is the law of conservation. It tells us that matter and energy cannot be created or destroyed in any system. Its quantity must remain the same. I say

11

matter and energy because we know they are equivalent. While they are different to us in our daily experiences, they can be converted into each other, and the formula $E=mc^2$ tells us the conversion rate. Because it's conserved, we know there must be a fixed quantity of it in the universe. This is our largest system in which every object has a relationship with every other object. They all influence each other by gravity.

This law must be as precise as any of our other laws. And I believe it can be expected to be absolutely precise as the great majority of our scientific predictions rely on quantity being constant. They rely on total matter and energy remaining equal. We symbolize the law of conservation with the equals sign and use it in our scientific formulas. It may be easy to forget the importance of conservation and believe it only comes up once in a while. But it really is front and center in science.

Returning to the uncertainty principle with this law, if the amount of matter and energy of the universe is precise, then its size or the area where the matter and energy may be found becomes infinite. Now we have a problem of the other extreme. Instead of infinitely small, this makes the universe infinitely large. But there's one more factor we need to put into place, and it will solve this problem. There must be more than one object in the universe.

This requirement comes out of quantum theory. It has to do with minimum quantities of everything in our world leading to a maximum particle size. If our universe was of small enough mass, then one particle may be able to contain it. But it's not. The universe is not, and never could have been, one indivisible piece of energy or matter because it has too much mass and energy.

By dividing a precise amount of energy and matter into more than one piece, we have some certainty of the quantity of each piece, but not absolute. We know each one is greater than zero, but less than the maximum energy a particle could have. This gives us some certainty of location, or size. Now we can begin to determine what our universe first looked like when it came into existence.

When we refer to subatomic particles, which are the smallest pieces of energy or matter in the world, size has a special meaning. It's different than we may expect but can be imagined as comparable to some aspects of our world. For example, if you walk a dog on a leash, you could say the location of the dog is where you are plus or minus the length of the leash in any direction. The length of the leash is the uncertainty of the location of the dog. You could also call it the size of the dog's area. Fundamental particles don't so much have a size as they have an effective size, or area where they may be found. So any particular particle, such as an electron or proton, could be described as being different sizes. You could say its size varies according to how much leash you give it. The more you shorten the leash, the more energetic the particle is, and the more area you give it, the calmer it becomes. This is basically how the uncertainty principle works. Uncertainty of location generally equates to size, and uncertainty of momentum is total mass (or mass equivalency of energy) x speed.

It's only when we're considering fundamental particles that this counter-intuitive behavior applies (massive objects being small and light ones being big). In our daily lives, we interact with objects made of lots of particles. Everything big enough for us to see is made of many fundamental particles. And the benefit of this type of behavior among them is the great precision of our macroscopic world. The uncertainty principle gives size to the particles which make up the things around us. The great quantity of particles which everything is made of, allows us to be sure of what they are and how they behave. At our large scale, this inverse relationship between size and mass is seen as density.

Uncertainty

In 1927, Werner Heisenberg presented this uncertainty principle. It told us what things we are not able to know. And in the process has allowed us to know things we didn't, since the other side of uncertainty is certainty. Properties are partnered. The more certain we are of one property, the more uncertain we are of its companion property. The smaller one quantity becomes, the bigger the other becomes.

This describes the inverse relationship between a particle's momentum (mass x speed) and its location. And this inverse relationship can be used to make very accurate predictions.

For a group of 226 particles confined to a 9.1×10^{-15} meter radius (a location with a precision of plus or minus 9.1×10^{-15} meters or 0.0000000000000091 meters) we're able to calculate that one of these particles has a 50% chance of escaping in 1620 years. This is due to its uncertainty of momentum, or location. This is the atomic nucleus (the center part of an atom where most of its mass is) of radioactive radium, and 1620 years is known as its half-life.

This may seem like a lot of particles confined to a small space, but that volume measured in Plancks (our smallest quantity of distance) is 7.476×10^{62} cubic Plancks, or 747.6 trillion trillion trillion trillion trillion.

The uncertainty principle tells us that the more we increase the concentration of particles, the more we increase density of particle count, and the more certain the particles are to escape.

Now imagine the certainty of escape when all the particles of the universe are located in no space at all. It would have to explode instantaneously. But that's not exactly how uncertainty works. That's not how radioactive elements, those with dense nuclei, such as radium, emit part of their mass in the form of energy.

Tunneling

The process of a particle leaving a nucleus is called tunneling, or quantum tunneling. In this maneuver a particle doesn't move out of the nucleus; it ceases to exist inside the influence of the strong force

14

which holds it there, and proceeds to exist outside of its reach. The strong nuclear force is very strong but only extends a short distance. This way a particle doesn't have to overcome any force in order to escape.

Tunneling may sound like a rare and unusual occurrence, but I believe it's very common. I believe it's the way all movement occurs.

According to my own theory of movement, everything travels by leaving one place and time, and appearing at another place and time. It doesn't pass through the space or time in-between. Of course, this does happen very rapidly. What this means for tunneling is that it's just ordinary movement. And being ordinary movement also means it's subject to the speed limit of light speed when traveling a distance. So I suggest tunneling may not always be instantaneous, as is popularly believed.

Quantum tunneling is also said to be how atomic nuclei bouncing around in stars get past each other's electrical resistance barriers in order to bond into larger nuclei. Since they are all positively charged, combining them is like pushing two magnetic north poles together. Inside a star, these nuclei bounce off of each other with great speed, which we refer to as heat from pressure. At high speeds, such as in a particle accelerator, two particles may have enough momentum to allow them to occupy a precise location according to the uncertainty principle. This is nuclear decay in reverse. The particles are passing into the space past the electrical resistance barrier to be bound by the strong nuclear force. And this applies to all extreme confined situations, such as those created by gravity.

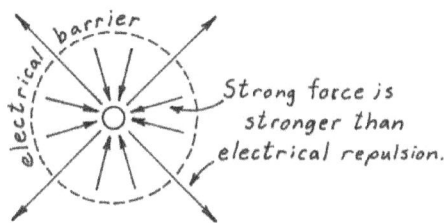

Gravity

Gravity is the strongest force in our world disguised as the weakest. Like Superman as Clark Kent, gravity in our everyday world is mild. It doesn't hold atomic nuclei together like the strong nuclear force, or bind electrons to their

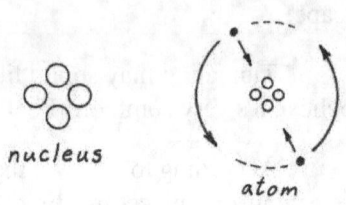

nucleus

atom

atoms like the electromagnetic force does. It only pulls together objects gently, until it gets enough mass in a small enough space, then it can become very strong.

Force = Energy / Distance

Force is not energy. It's defined as energy divided by distance. This tells us that it's an expression involving energy. It is an effect of energy which is distance, or size, related. This means it's also not mass, but is an effect of mass. And it's in fact not an object at all if it's not energy or mass. This will be an important concept as we proceed.

The force of gravity decreases exponentially as we leave a massive object like earth. Even greater is when we approach an object with more mass than, or which is more confined than the earth. A good example of an extremely dense object, or collection of particles, is a neutron star. One cubic centimeter of its material weighs 100 million tons. Neutron stars form because of gravity, and as they form the gravitational force at their surface increases as their size decreases.

Imagine the gravitational pull of all of the mass of the universe contained in no space at all, a singularity. Having no volume its density would be infinite, and so would its gravitational pull. (Mysteriously, it would be stronger than all of the energy it's made of.) No force could pull it apart. So no force does. The objects of the universe leave the singularity by quantum uncertainty. These types of situations show gravity to be the strongest of all forces since it has no obvious maximum limit like the strong force does.

Quantum uncertainty doesn't pull anything apart. It uses no force, so it's able to dismantle anything. It could be said to go around force.

The infinite certainty of location which a singularity has, gives it infinite momentum. But zero size also gives it infinite gravitational force to counter the infinite momentum. We have a standoff of infinities.

In the case of our universe, we know that it does have a definite amount of matter and energy, and that gives it no definite size if it was to be one object.

This tells us that the Big Bang was initiated, and the universe came into existence because the law of conservation was put into place and the total mass and energy of it became certain. The quantization of our world, then, required mass and energy to be divided among particles. And these particles required room.

The last physical law written was the law of conservation.

If we had all of the universe located in one place of no size, where uncertainty will not allow it to remain, it doesn't overcome the infinite force which confines it. It ceases to exist there. (In fact it never did exist there because it couldn't.) And it proceeds to exist at multiple other locations. I believe these locations can be determined by probability. Since the particles' size, momentum, and other properties are determined by probability, their location should be, too.

Probable Distribution

Probable distribution was explained to us by Karl Friedrich Gauss (1777 - 1855), and it has a definite form. We call it Gaussian distribution and describe it with the Gaussian curve, or bell curve.

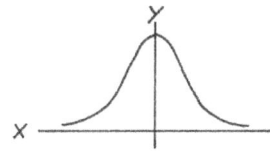

17

This distribution is described mathematically by

$$y = e^{-x^2}$$

in which y is the vertical axis, or the probability of finding an object at that location on the x-axis, and e is Euler's constant of 2.71828...

In this curve diagram, the x-axis represents all possible locations. So, the curve tells us the concentration of objects at any location.

This, then, should describe the initial arrangement of the universe.

If it originated as a singularity, then it began its existence, or came into space and time, as a cloud of particles distributed according to Gauss. Dense in the center, sparse at the extremities, and potentially across all of space because the low ends of this curve extend indefinitely.

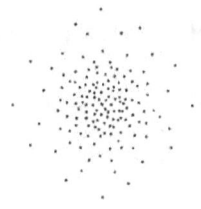

This fits with Georges Lemaitre's 1927 suggestion that the universe began as a cosmic egg.

What determines the span of the bulk of the particles is their probability of remaining at that location. It seems that the particles should begin their existence with enough room that they won't immediately tunnel out of it.

We can find this using Heisenberg's formula for uncertainty in the form of:

$$\Delta p \Delta q = {}^h\!/_{2\pi}$$

18

where

Δp = momentum (or mass in kg x speed in m/s),

Δq = radius in meters,

h = Planck's constant of 6.626 x 10^{-34} Joule-second.

The bell curve shows us the distribution. Heisenberg's formula tells us the size of the particles we're distributing, and therefore the size of the bell.

Size of the Initial Universe

It is believed that there are 10^{87} particles in the universe and that its total mass is 1.73 x 10^{53} kg. This mass includes dark matter and dark energy, which are said to be 95% of all matter and energy.

I'll exclude them from our calculations because I believe they're unnecessary concepts, and will explain why later. So, for this calculation I'll say that the total mass and energy of the universe is 8.65 x 10^{51} kg (the 5% which is ordinary matter and energy).

Using this figure, we can estimate each particle to be 8.65 x 10^{-36} kg. We can also give each one a random speed between zero and light speed. This makes their average speed half of light speed, or 150,000,000 m/s. Their average momentum (mass x speed), then, is 1.297 x 10^{-27} kgm/s.

Minimum Number of Particles

Earlier I mentioned that our universe could not have ever been one individual piece of matter or energy. This is because our physical laws give us size limits. The most energy, or mass, one particle may have is limited by the smallest it may be. We learned this as part of quantum theory.

Discovering that there are smallest measures for everything led to the finding that there are largest measures also. The first quantum discovery was made by Max Planck in 1900 when he found that there is a smallest amount of energy. Today it is known to be

6.626 x 10^{-34} Joule-seconds, and is called Planck's constant and symbolized by h. His original finding was the relationship between energy and the frequency of energy waves ($E = hf$, where E is energy, h is the minimum quantity, and f is frequency). From this, the smallest measure of distance was 1.616255 x 10^{-35} meters, and is referred to as the Planck length.

The smallest measure of distance gives us a maximum amount of energy since energy and wavelength (area where a particle may be found) are inversely related. This is due to the inverse relationship between wavelength and frequency ($f = c/\lambda$, where c is the speed of light and λ is wavelength). We can determine the maximum energy by entering the minimum distance into Planck's formula as wavelength:

$$E = {}^{hc}\!/_\lambda$$

where

E = energy in Joules,

c = the speed of light as 299,792,458 m/s,

λ = wavelength in meters, which we can designate as the Planck length.

What we learned is that the most energy a particle can have is equivalent to 1.37 x 10^{-8} kg of mass (according to $E = mc^2$ using the speed of light in m/s).

This tells us that for the amount of mass that our universe has, it must be distributed among at least 6 x 10^{59} particles. Since fundamental particles can combine or separate, it's possible that our universe began with as few as this number of particles and they divided later. But for now, I'll proceed with our larger, lighter particles and we'll see how big the cosmic egg could have been.

Heisenberg's formula tells us that each has an uncertainty of location of 8.138 x 10^{-8} m. The volume each would occupy, then, is 2.822 x 10^{-22} m^3. (The volume of a sphere is 4/3 π r^3.)

If we distribute 10^{87} particles of this size according to probability, using the formula of

$$y = e^{-x^2}$$

we find a distribution which looks like this:

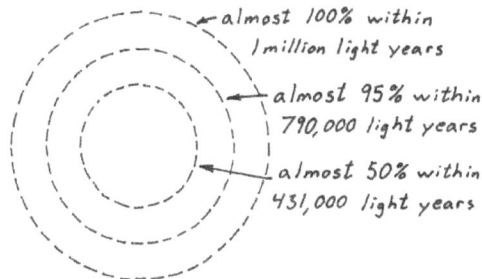

Using our maximum particle mass of 1.37 x 10^{-8} kg, each has a radius of just 3.83 x 10^{-27} m and volume of 2.357 x 10^{-79} m³. Distributing 6 x 10^{59} of these according to probability puts almost 100% of them within 8.093 x 10^{-7} m, or about a micron. That's the universe, in a very small speck of dust.

Our universe may have come into existence as a cloud of particles no bigger than 1 million light years in radius, and no smaller than a speck of dust.

Process of Distribution

Previously I suggested that quantum tunneling is normal movement, and that it's how these particles get out of a singularity and into their initial locations in space. For this reason they could not have all appeared at once. If nothing can travel faster than light then certainly the most central particles appeared first. And those which may have initially come into space a million light years out, would have taken a million years to travel that distance.

By this reasoning the whole universe could not have come into existence instantaneously. And it wouldn't spend a million years entering space, beginning with the center particles and working outward. They won't wait for the outer particles to appear before they begin interacting. The particles would appear at the center first and push away from each other with complex movements making room for more. They would appear to flow out of nothingness into space like gas entering from an invisible spigot.

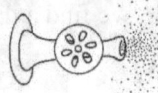

There's an obstacle to making the universe bigger from here. That is gravitational force. If the cloud of matter has enough mass in a small enough area, it won't be able to grow because gravity will be too strong.

Having come into existence distributed according to probability, its center will be densest and most at risk of needing to be re-distributed. Since the particles are active, some of them will come together and create uneven densities. Some areas certainly will become too dense and require re-distribution.

The result of the first distribution of matter in space is a series of re-distributions continuing to occur until the majority of particles can move about freely.

The universe didn't begin with just one explosion, but one big explosion followed by many smaller re-explosions of that same matter. This must be how the universe attained size. And this is necessary because, as we'll see later, no force could have caused this growth.

Chapter 3 - Gravitational Interactions

Once in place each particle has not only a location but also a speed and direction determined by probability. Location is based on a probability curve, so the particles are most likely to be concentrated at the location of their most recent redistribution, or explosion. Once these explosions subside, they'll begin a more familiar interaction with all speeds and directions being equally possible.

Similar to how light (in particle form we call it photons) is emitted by spontaneous emission, probability does not favor any direction over any other. This is why light bulbs illuminate equally in all directions. The initial particles of our universe should also go every direction and travel at every possible speed, from zero to light speed.

Now that we have a lot going on, we'll focus on gravitational interactions. Just as I pointed out previously, it would take time for all of these particles to appear in space, so their gravity waves should also take time to get into place. Gravity is known to travel only at the speed of light. This means as these particles first come into existence their neighbors don't immediately notice.

One-on-one, particles aren't likely to respond to their gravitational attraction. Each has far too little mass for its size to have any noticeable attraction to each other. Particles generally bounce off of each other before they get close enough for that. But as a whole, the universe is drawn to collapse by gravity. Intense interactions among particles which average 1/2 of light speed provide an opposing pressure.

Any particles which are pure energy may occupy the same location and pass through each other. Those which are matter must either bounce off of each other, combine, or break apart.

Rotation

Every dense accumulation of particles must rotate. As energetic particles continually collide in their attempt to escape, some agreement will be made as to how to relieve some of the pressure.

They must move, and they want to move straight. If they can't, their best alternative is to rotate. By chance, one direction will gain preference. Only a slight tendency is needed to cause the particles to take advantage of that.

Clumps will rotate. Even the initial cloud will rotate. The whole universe, having begun as one energetic cloud, must rotate in order to relieve pressure. Like an energetic animal confined to a small area, it will travel in a circle because that is the straightest path to expressing its energy. And energy must always move.

But rotation occurs in 2 dimensions which is why it's so common in our atmosphere. Our atmosphere is much wider than it is high. Matter which can arrange itself as a disk can easily rotate. Spherical objects have more difficulty. They may rotate more haphazardly in various individual relationships.

The Spiral

Particles, and even larger objects made of groups of particles, will either collide as they travel or pass by each other. If they pass by, their paths will be curved by gravitational attraction. Their curve will tighten as they near each other and gravity becomes greater. And they'll get wider after they pass as gravity is less.

Every gravitational interaction creates a spiral path for each object involved. This is because gravitational attraction increases as they approach, and decreases as they depart. Since gravity is never gone, there is always some curve.

It seems to be popularly believed that a circular or elliptical orbit is another possible relationship, a balance of gravitational and centrifugal (outward) force. This would mean the spiral could be closed to form a continuous loop. This is a useful concept for many purposes, so I'll proceed to use it. But we'll see later how another

24

factor could prevent stable orbits from occurring. Although it's often slight, we'll see how orbits constantly change.

If the particles have clear paths they continue on their way. If they collide with anything or encounter multiple gravitational influences then they may slow. This allows objects to pull closer to each other. The result is that denser identifiable objects like planets and stars are formed.

Since most everything which passes by each other without colliding proceeds on a spiral path, all objects will rotate around everything they interact with. This leads to free bodies developing orbits.

Both the vibrational energy of compact bodies seeking the relief of pressure and the spiral maneuver of objects which are able to pass by each other cause rotation. The result is that planets, stars, solar systems, and galaxies all rotate.

Spinning and orbiting are almost the same, and for now we'll consider each to be an indefinite circular path. Indefinite because of inertia. Circular because it's the straightest path allowed in an orbit.

Stars are considered to orbit a galaxy because there's a lot of visible space between them. Planets and stars are considered to spin. Although they too, are made of lots of objects traveling together, we can't see the space between them. We also know that they are relatively closer and have more complicated interactions. The components of a spinning object bounce off each other. Objects in orbit don't. More precisely, at short distances we have effects from electrical forces in addition to gravity. At great distances, it's just gravity.

It seems that the initial universe was held together by gravity, but it couldn't collapse because of energetic collisions and redistribution by the uncertainty principle. Additionally, rotation prevents collapse.

I believe what causes our universe to evolve from its early state to what we have now, including its expansion, is gravity. So we'll look closely at what gravity is. We'll begin with Newton's discoveries.

Force Of Gravity

The force of gravity is based on the mass (or mass equivalency) of the objects attracted and the distance between them.

Our formula for calculating it was given to us by Isaac Newton (1642 - 1727) and I believe it tells us of the distribution of the energy of an object.

Every object is made of energy. It is the underlying substance of matter. Some of it may be expressed as matter and some as movement or kinetic energy. But all of it is always a factor in gravity, so far as we have been able to determine.

Newton's formula for the force of gravity is $F = GMm/r^2$, and it is a version of his basic formula for force ($F = ma$) modified to specify acceleration caused by gravity. The G/r^2 portion is the acceleration caused by gravity according to the inverse square law. It tells us the gravitational force of a base unit of matter at every distance from an object. And the Mm tells us how the mass of both objects is involved. M is the mass of one object and m is the mass of the other.

Inverse Square Law

Everything which emanates in 3 dimensions of space, such as gravity, disperses in 2 of them. This is because the first dimension allows for travel and the others allow spread.

Water flowing through a hose is traveling one direction; a line is one dimension. The pressure, or force, coming out of the end of the hose is the same as its pressure going in. In one dimension there is no decrease because there is no spread.

Place the hose nozzle on a flat surface, such as the hood of a car and the pressure of the water will be reduced in direct proportion to its distance from the source. The water pressure 2 inches ($r = 2$) from the nozzle will be 1/2 of what it is 1 inch from the nozzle. At 3 inches ($r = 3$) it will be 1/3 of its pressure at 1 inch. A force spread across 2 dimensions decreases in direct proportion to distance ($1/r^1$).

26

Next, place a ball on the nozzle which is made of a fine screen to allow water to spray in all 3 dimensions equally (up/down, left/right, forward/backward). Now the pressure is reduced according to $1/r^2$. At 2 inches ($r = 2$) away from the nozzle, pressure is $1/2^2$, or 1/4, of pressure at 1 inch. At 3 inches ($r = 3$), pressure is $1/3^2$, or 1/9, of pressure at 1 inch. This is the inverse square law.

This is why the gravitational constant (G) is divided by distance (r) squared (G/r^2). If we lived in 4 spatial dimensions, then spread would occur in 3 of them and we would apply an inverse cube law (G/r^3).

Masses

The two $m's$ are the mass of each object involved in the attraction, since there is no attraction if there is only one object. Gravitational force depends proportionally on both masses individually. This is why the two masses are multiplied by each other, and by G, not added. If they were added, the smaller mass would have little effect.

The gravitational constant has unusual looking units. In order to understand why, we'll develop the formula for gravity and see how they make sense.

Creating The Formula

To see how this formula can be created we'll begin with $F = ma$ since we know it already expresses an equivalency for force. Since acceleration (a) is distance per time squared, we'll expand the formula to

$$F = m \cdot \frac{r}{t^2}$$

where

F = force,
m = mass,
r = distance,
t = time.

27

To put more factors into it we'll need to either perform the same function to both sides, or effectively multiply one side by 1 in order to maintain equivalency.

Newton found that the force of gravity is inversely related to the distance between the objects squared. We need to divide by r^2 without disturbing the existing equivalency. We can do this safely by multiplying one side by r^2/r^2 because it's equal to 1:

$$F = m \cdot \frac{r}{t^2} \cdot \frac{r^2}{r^2}$$

The force of gravity is directly proportional to the mass of both objects involved. So we need to multiply the whole thing by Mm. We'll do this by multiplying by Mm/Mm to maintain our equivalency:

$$F = m \cdot \frac{r}{t^2} \cdot \frac{r^2}{r^2} \cdot \frac{Mm}{Mm}$$

(We have some extra units out of necessity, but they won't have quantities, so they won't affect the math.)

The gravitational force is a part of the mass of an object for its distance. And proportionally it's the same for all objects. It is a constant. The constant will be made of our left-over units. We call it G. Putting this into our formula makes our force the force of gravity. We write out our constant and the variables we're using as

$$F = \frac{GMm}{r^2}.$$

The rest of our units get packaged as the units of G:

$$G = quantity \cdot \frac{mr}{t^2} \cdot \frac{r^2}{m^2}$$

Measured in kilograms, meters, and seconds we've found this constant to be 6.672×10^{-11} kgm/s^2 · m^2/kg^2. Today 1 kgm/s^2 is known as a Newton. So G is commonly written as 6.672×10^{-11} Nm^2/kg^2.

Measurements of the gravitational constant vary, so it's not known with great precision. The amount ranges from 6.6699 x 10^{-11} to 6.6745 x 10^{-11}, a 0.07% variation. In 1986, Ephraim Fischbach proposed a 5th force to explain the inconsistency. Paul Dirac suggested that G may decrease with time. However, conflicting with these noted variations, G is also believed to be related to other physical constants which are more precisely known:

$$Planck\ length = \sqrt{\frac{hG}{2\pi c^3}}$$

Variations in measurements of gravitational force seem to indicate that there's a factor not being considered.

But for now, let's look at what the force of gravity does.

Chapter 4 – Condensing Matter

Objects are believed to emit energy in the form of waves which reach out in all directions and to all distances, affecting every other object. And those objects do the same.

The waves are involved in pulling together the objects they're a part of and we call the waves, or the particle aspect of them, gravitons.

All objects, all matter and all energy, is always trying to come together. While matter particles are generally not allowed to occupy the same location, their waves can. These energy waves bring their matter close. When enough matter gathers closely enough the force of gravity can cause very dense materials to form.

Planets

As particles gather and pack together, they are forced to orient themselves efficiently. Carbon atoms find that they fit together more closely in the crystal structure of graphite.

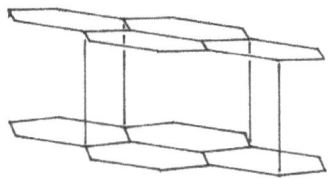

With more pressure they are forced into the tighter form of a diamond.

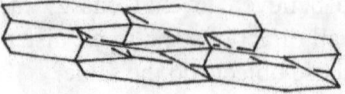

All types of minerals and other rocks form by fitting together the particles which are available. We find this done by planets because they have enough gravity to provide the needed pressure.

The vibrational energy of each particle, its heat energy, is how they move around to find their fit, and also gather together identical particles for a better fit.

Elements themselves (like carbon) are made under higher pressure.

Stars

With increased mass comes increased gravity and greater pressure, which results in increased vibration (what we know as heat). With enough heat, though, atoms can't hold together electrically anymore. Their collisions become so strong that bonds between the positively charged nuclei and negatively charged electrons which orbit them break.

Actually, the impacts transfer energy to the electrons in the form of photons which cause them to move to higher orbits until there are no more orbits left.

Since electrons are low mass particles which form a hollow shell around their atom's nucleus, they're the first layer of matter to break apart. The nucleus is so much smaller than the shell that an atom is as hollow as a sports stadium containing nothing but a grain of rice.

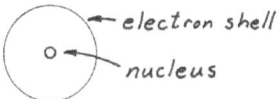

Breaking this shell allows collapse.

Gravity causes pressure which results in high speed collisions breaking atoms apart so that they may collapse into a higher energy mix of free nuclei and electrons. Some protons and electrons will join to form neutrons.

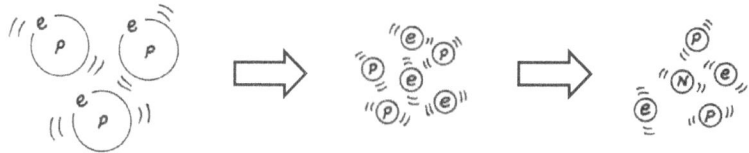

Protons can join protons, too. Although they are positively charged, they can overcome their electrical resistance to each other with enough speed. They just need to get within range of the strong force which can bind them. This is like forcing two magnets together with like poles and tying them that way. With enough matter together, such as in a star, a strong enough gravitational force is created to squeeze the core of it together and make protons merge (or fuse) into large nuclei so that they take up less space.

They are bound by the strong nuclear force which, as waves or particles, is known as gluons.

These larger nuclei are new elements made of various combinations of protons and neutrons. This is how we get the different elements. The neutrons give atoms additional weight, but the protons provide the positive charge which then gets balanced by an equal number of electrons to give the atom its properties.

All the various elements are formed by breaking apart nuclei and re-assembling them in different arrangements.

In the merger process, referred to as nuclear fusion, lots of energy is released. This energy originates as gamma rays (very high energy light) but escapes as visible light after being absorbed and re-emitted for tens of thousands of years on its way out of this massive ball of fusing nuclei. This ball is a star and temperatures inside can reach 50-100 million °C.

With enough mass, enough gravity when fusion slows, a star could collapse further. This could result in an explosion sending new elements throughout space to eventually form planets or other stars. Or, the star could hold together and remain collapsed.

Neutron Star

After burning most of its fusible material, forcing together the small particles it's capable of, the gravity which had always been holding it together is finally able to make the star smaller. This is because fusion has slowed and enough of the energy released by it has escaped. Its outward pressure has reduced and the star is able to cool. But it doesn't. Instead, it collapses further because, while the pressure keeping it inflated is gone, gravity remains.

Various collapses of stars are possible, each initiating a higher temperature fusion process to build larger nuclei for heavier elements. But one collapse forms the ultimate atomic nucleus. It's the one which

34

has enough gravity to force all of its electrons and protons to merge into neutrons and all neutrons to bond into one big neutrally charged atomic nucleus. Because of its neutral charge, electrons can't establish orbits to make it an atom, but it's still a nucleus. It's a neutron star.

The Schwarzschild Radius

There's another stage of collapse which matter is capable of, and it may be the final stage. It can occur because atomic nuclei have empty space in them, too. While neutrons are packed together with each other tightly, neutrons (and protons) have empty space inside them.

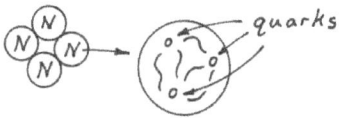

Inside of each neutron's, or proton's, 10^{-15} meter diameter, are 3 tiny quarks and lots of energy which hold them together (more gluons). This object, too, can collapse, but not necessarily completely. The degree of collapse, I believe, is controlled by the probability of escape.

This is one way which a well-known dark star can be created. They were speculated about over 200 years ago and later found to exist.

In 1783, John Michell announced that he believed it is possible for there to be an environment in which the force of gravity is inescapable.

Then, the year after Einstein first presented his general theory of relativity and the year it became published, this concept was explained. Since the general theory of relativity is very much about gravity, it was able to lead Karl Schwarzschild (1873-1916) to a solution. He notified Einstein that an amount of 2 times mass multiplied by the gravitational constant and divided by the speed of light squared tells us the radius within which nothing can escape. Its inescapability is based on the limiting factor of light speed. Since nothing can travel faster than light, this meant that nothing can leave. He had written the formula for an object which John Wheeler later named a "black hole":

$$R_S = \frac{2Gm}{c^2}$$

where
R_S = the Schwarzschild radius in meters,
G = the gravitational constant of 6.672 x 10^{-11} Nm^2/kg^2,
m = mass in kilograms,
c = the speed of light as 299,792,458 m/s.

A black hole can be a star which is so compact that its own light can't escape. Our sun would have to be less than 6 km in diameter to be a black hole. Right now it's 1,391,900 km in diameter. But a black hole doesn't have to be a star. It only needs to be a collection of matter or energy within a certain size which causes gravity to be inescapable by anything, even light. That's why they're called black. We can't see inside them because light can't get out.

According to Schwarzschild's formula, small black holes need to be very dense but large ones don't. The larger a black hole is, the less dense it will be overall.

If we eliminated the sun and all the planets, meteorites, and dust from our solar system and filled it with air at our normal atmospheric pressure, just as we breath, our solar system would have inescapable gravity and be a black hole. Of course this mass of air wouldn't remain evenly distributed. It would interact gravitationally and eventually form planets and maybe even a star.

While a large black hole has less overall density, the mass inside it may exist as very dense objects with lots of empty space between them.

Deriving the Formula

The Schwarzschild formula was derived by applying the very

difficult mathematics of Einstein's general relativity. It's known to be one of only two precise solutions ever found. For that reason, black holes have been viewed as relativistic phenomena, extreme warping of space and time in accordance with Einstein's model of the universe. However, we can find an alternative meaning to this formula if we develop it in another way. The method we use to create a formula can reveal what the formula is doing, and therefore its meaning.

I find the formula can be created using a few simple concepts. We can derive the Schwarzschild formula for the radius of this inescapable environment by combining Newton's formula for gravity ($F = GMm/r^2$), which was published in 1687, with Galileo's formula for distance traveled while accelerating ($r = 1/2 \ at^2$), written around 1609. (This is believed to be the first precise scientific law.) We can use this formula because it describes our attempted escape. We're traveling a distance while being accelerated against. If we're being accelerated against with a certain force, then it will tell us how far we can go.

We begin by adjusting the form of each.

Because force is mass x acceleration or mass x speed/time, then $F = GMm/r^2$ becomes $mv/t = GMm/r^2$. And $r = 1/2 \ at^2$ becomes

$$r = \frac{1}{2} \cdot \frac{v}{t} \cdot t^2 \ \text{ or } \ r = \frac{1}{2} \ vt$$

once we cancel one of the t's. Our two formulas, now, are

$$\frac{mv}{t} = \frac{GMm}{r^2} \ \text{ and } \ r = \frac{1}{2} \ vt \ .$$

We can replace one of Newton's r's with $1/2 \ vt$:

$$\frac{mv}{t} = \frac{GMm}{r \cdot 1/2 \ vt}$$

Multiply both sides by $1/2 \ vt$ and divide both sides by m, then cancel the t's on the left, for

$$\frac{1}{2}\,v^2 = \frac{GM}{r} \quad \text{or} \quad \frac{v^2}{2} = \frac{GM}{r}.$$

We can solve for the escape velocity from a gravity of *GM* at *r* distance by multiplying both sides by 2, then taking the square root of both sides:

$$v = \sqrt{\frac{2GM}{r}}$$

This tells us how fast we need to start out going to travel against that force for that distance. Since nothing can travel faster than light, light speed is the fastest possible escape speed.

If we square both sides, then divide each by v^2 and multiply by *r*, we have

$$r = \frac{2GM}{v^2}.$$

This tells us the distance we're able to escape to if we travel at a speed of *v*, against that amount of gravity. Using our maximum possible speed, light speed *(c)*, *r* becomes our maximum possible limit of escape:

$$R_s = \frac{2GM}{c^2}$$

This is the Schwarzschild radius, the formula for the size, or what is known as the event horizon, of a black hole.

Though the formulas I used are over 300 years old, the size of a black hole could not have been determined until the speed of light was established as the maximum speed of travel. Einstein determined this as part of his special theory of relativity in 1905. He expressed this in his paper on the subject when he wrote, "Velocities greater than that of light have... no possibility of existence." This maximum possible escape speed was the key piece needed for defining what an inescapable environment is.

Seeing how this formula is constructed, that it's based on a constant deceleration over a distance, reveals a few important aspects of it. It describes leaving the center of a black hole at an instantaneous speed and with no further propulsion. If the force of gravity is that of at R_S and it remains constant for the distance of R_S, the object will be unable to go any further than that distance even if its initial speed is that of light.

Notice that the gravitational force used is that at R_S. Force may be different at various locations inside a black hole. This tells us that the distance attainable in an escape attempt may vary depending on the density or arrangement of matter within a black hole. It may also depend on where the object departs from and whether it has propulsion, or some source of acceleration outward.

The Schwarzschild radius is not a hard line surrounding a black hole. What we refer to as an event horizon may not be inescapable. One might, then, suggest that black holes don't exist. But they do. They exist because they've been defined by the Schwarzschild formula and scenarios fitting its description exist. They're just not as we have believed they are. With this in mind, let's look closer.

Chapter 5 - Black Holes

Since we have a definition of what black holes are, provided by the Schwarzschild formula, we may begin to explore them and learn what they do.

Outside

If we were to approach a black hole directly we would accelerate and enter it. And we may not be able to leave. But if we approached indirectly, we could establish an orbit around it just like we would around any other gravitational body. A black hole can be orbited as easily as a star or galaxy can be. In fact, all galaxies are believed to have a black hole at their center which their stars orbit. The difference we might notice between a black hole and a planet or star, is how close we're able to get to its center and therefore how much gravity we're able to feel. So, around a black hole, closer, faster orbits may be established than around a star.

If our sun was to somehow become a black hole, it would be much smaller than it is now, but our orbit around it would be unchanged. The sun's gravitational force wouldn't change at any distance except that we would be able to get closer and experience the greater force within its old radius.

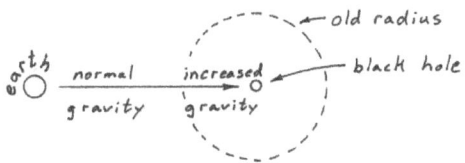

Just as objects commonly orbit stars, there are usually objects circling around black holes. Sometimes it's dust. Larger black holes can have a star orbiting it. Very large black holes have an entire galaxy orbiting them, such as our own galaxy is.

Inside

We know that the perimeter of a black hole is at a distance from the center which light is unable to travel under a certain amount of attraction. Some believe that inside this distance gravity continues to get stronger. This would mean objects continue to fall to its center while continuing to accelerate. If so, then that's where all of its mass and energy must be, becoming smaller and smaller. The center would be a singularity. And joining it would be the destiny of anything which crosses the perimeter.

However, we saw earlier how a singularity gets dismantled immediately by quantum uncertainty when it's composed of multiple objects of limited total energy. This would describe all black holes we know of. For this reason they could not have a singularity in them.

If an object did attempt to collapse into a singularity, as we tried to do with the initial universe, it should do as the universe must have done. The collapsed object would have to explode due to the uncertainty principle and re-distribute its contents according to probability. There would be a cosmic egg formed. I believe the contents of the black hole would seek pressure relief by way of rotation. The contents would either rotate fast enough to avoid collapse or explode and re-explode until it did. The contents of any dense accumulation would seek some degree of distribution which is acceptable to the uncertainty principle. Each particle would seek elbow room.

This process applies to matter. Any energy (light) within a black hole will be attracted to each other, but will pass through. Its destiny should be to establish an orbit inside the environment at the speed of light. We'll learn more about this later.

I believe we now have an object stable enough that we can proceed to imagine entering it and learn about its gravity. We can imagine that we're entering an object with some distribution of matter. This is similar to a gravitational environment we're all familiar with, the earth. The earth is actually most dense at its core, but we can imagine that it's an even density.

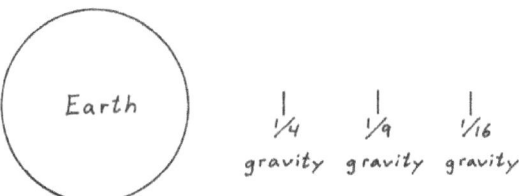

The force of gravity on the surface of the earth can be calculated using Newton's formula. From the surface, we know that gravity decreases as we leave earth. The force of gravity leaving is inversely proportional to our distance from its center squared.

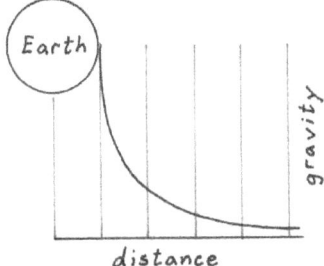

This suggests that below the surface of the earth gravity would become stronger by the same law. But it doesn't.

Standing on the surface, all of its mass is below us, so we feel all of it. But if we dug a hole into the earth there would be mass on the sides of us, and even above us as we dug deeper. This mass would attract us to our sides and upward as the lesser mass below us attracts us less. Some of the mass which pulled us down as we stood on the surface would pull us in all different directions as we descended. We would find that our weight decreases the further down we go.

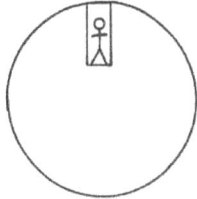

Once we reach the center, any amount of pull downward, or to the side, would be balanced by an equal pull upward or to the other side. We would weigh nothing.

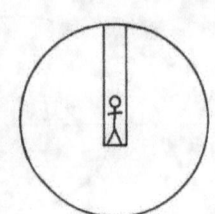

Gravity pulls us in any direction according to the amount of mass (or energy) in that direction, and our distance from it.

Newton's formula for gravity outside of a gravitating body tells us the force of gravity as we stand on the surface of the earth. This is because outside of a body, gravity is measured from its center.

Inside a body, we still measure from the center of gravity, but we need to account for the mass which we are inside of, as well as what we're outside of. Looking toward the center of gravity, some of the mass in that direction will be canceled by the mass which is behind us. And mass on one side will be canceled by mass on the other side.

If we sent a probe to a star, it would experience the same thing. If the probe were to drill to the center of the star, it would find no gravitational pull there. Interestingly, that's where all the nuclear fusion takes place, and the fusion is powered by gravity. However, it's the gravity from the outer parts of the star which pull them inward and create pressure. And the pressure is what causes the particles to have high energy to collide and fuse. Nuclear fusion inside a star isn't caused by gravity directly, but by the pressure from surrounding material being pulled together by gravity.

Now let's go back to our star which is collapsing to form a black hole. Once part of the star has collapsed so that the mass inside a radius can create a strong enough gravitational force, we have a black hole. This occurs as the surface reaches R_S. If more matter or energy enters this radius, the radius expands accordingly. This doesn't cause the object to expand, nor does it cause a relief of pressure.

This means its volume grows faster, and the black hole loses overall density as it grows, only as measured by the Schwarzschild radius. But the object inside the radius can still continue to collapse. Gravity at its surface continues to become stronger as the object gets smaller, but the internal pressures grow also. With increasing difficulty of escape, even energy will be trapped, and the violence inside the star will intensify. It will become a contained explosion.

With this pressure, I believe it will rotate. And its contents will arrange themselves by orbit speed, slowest at its center, where gravity is least, and fastest at the outside. This is the opposite of typical orbit speeds because it's based on the amount of mass there is to orbit. Further out there's more mass within a radius, stronger gravitational pull, and therefore a faster orbit. It's the difference between being outside of a collection of matter and being inside it.

Black holes are very intense situations, so we'll look more at the workings of gravity to see what the future of them may be.

Chapter 6 – How Gravity Attracts

When Newton explained gravity to us, he told us that it involves both objects pulling on each other, not just one of them pulling on the other. So it's not just the earth that attracts us. We also attract the earth.

We know that both are involved because gravitational force varies among objects of different mass. In our daily lives we experience gravity as weight. While we consider weight to be a measure of mass, it's actually a measure of the gravitational force between an object and the earth, based on the mass of each. For an object of greater mass, there's a greater force pulling it toward earth.

You might think your mass is insignificant compared to the earth's, but in your personal relationship with it, your mass is just as important.

In Newton's formula of $F = GMm/r^2$ the masses are multiplied by each other because they each play a proportionally equal role. If either was halved or doubled, the total gravitational force would be also.

This formula indicates that the gravitational attraction we feel on earth is directly proportional to our own mass. A person of 10% greater mass feels a 10% greater pull toward the earth. This is what makes his weight 10% more, and why we can measure it on a scale.

But if he fell down, it might seem that he would also accelerate 10% faster. And that a cannon ball dropped from a tower would hit the ground long before a bullet, assuming each has the same shape and density. The greater force is countered by greater inertia, objects resisting acceleration.

There are two similar formulas we use for gravity. The rate of acceleration from falling is determined by GM/r^2, while an object's attraction to the earth is found by GMm/r^2.

GM/r^2 tells us that all objects accelerate to earth at 9.8 m/s^2.

47

GMm/r² gives us acceleration times a specific amount of mass. So the acceleration of an 80 kg person is 9.8 kgm/s² per kilogram, or 784 kgm/s² total.

Both formulas are variations of each other.

$$a = GM/r^2 \qquad F = GMm/r^2$$

Excluding factors such as wind resistance due to various shapes or densities, all objects fall at the same rate.

Communicating Gravity

The four forces of nature are believed to be communicated by particles. Both gravitating objects must be sending messengers to the other. In the case of gravity, gravitons are the particles being sent.

Gravitons are proven to travel as waves at the speed of light. Since only pure energy can travel that fast, they must be pure energy. What they do to create attraction is uncertain.

Potential Energy

A popular explanation for how gravity causes objects to accelerate toward each other, or a smaller one to accelerate toward a larger one, is that it converts potential energy into kinetic energy. The belief is that the energy used to give speed to a falling object is already a part of the object before it falls.

It's given to the object when it's moved away from a gravitational body. If I pick up a ball, the energy I use to accomplish that is given to the ball as potential energy. (Pushing the ball across the ground does not give it potential energy.) So if I pick up a ball of a certain mass, it then has that mass plus the mass equivalency of the energy of picking it up. When I drop it, as the ball falls, the potential energy is converted into kinetic energy. In this way the ball has more mass equivalency while it's in the air than it did before I picked it up. And it has even more mass equivalency the higher it's lifted.

This means that every object has an amount of potential

energy which is based on how far away it is from any gravitational source which it may eventually fall to. It may have potential energy determined by the size of the universe. In this way, as the universe expands all objects gain potential energy. This potential energy, then, must be provided by whatever mechanism expands the universe because that mechanism is "picking up" all objects. So an energy source is needed for expansion. This is the position many scientists are in today, trying to identify this mysterious, or dark, energy.

An expansive, or repulsive, force hasn't been found. In our world we have various wavelengths of energy from radio through visible colors to gamma rays. We also have matter, which is composed of energy. And all of this energy that we know of is attracted to all of the rest of it, not repelled. We don't know of an energy which is repulsive to all other energy and matter.

For energy to be repulsive it would have to be anti-energy or negative energy. We do have anti-matter. It's like ordinary matter but with opposite properties (opposite charge, opposite direction of spin, and traveling backwards in time). But we do not have anti-energy. We know that matter is made of ordinary energy. Anti-matter is made of ordinary energy, too, not anti-energy. Experiments prove this. When we combine a particle with its matching anti-particle (same mass and type but opposite properties) they annihilate and what remains is the energy of both particles. The energy left behind by this annihilation has the mass equivalency of the two original particles. There's no cancellation of energy. Both the particle and anti-particle were made of ordinary energy.

There's a reason why we have anti-matter but not anti-energy. Matter experiences time, energy does not. This is because energy travels at light speed and at that speed it's fully dilated, it has no experience of time or distance. Ordinary matter, the matter we're familiar with, experiences time in the same direction we do, what we consider forward. Anti-matter experiences it in the other direction. This is where it gets its opposite properties. So while we and the ordinary particle see an annihilation occur, the anti-particle experiences itself as being created. When we see an anti-particle created, it's experiencing annihilation.

I believe time is just a dimension. One direction of it is fundamentally no different than the other. We just experience time in this direction we call forward. And we can't experience it in the other direction because of the law of conservation. The reason is because we're already going in this forward direction.

If anything made of matter or energy could turn around in time and go the other way, then the past would have too much total matter and energy, and the future wouldn't have enough. It could, then, turn around again to go forward and save the future from being short, but the past would always have an excess. No matter how many times or when it turned around, the amount of matter and energy throughout time would never again be equal. I believe the law of conservation not only dictates that there is a specific amount of energy or matter for the universe, but that it be the same at all times. Once the universe begins there's no going back.

As we see with the dilation formula for travel, light cannot experience time because all experience of time and distance is given up at light speed. Light can't go backwards in time because it's not inside time; it's outside of it according to movement theory. Only matter can go backwards in time.

This doesn't mean there can't be anti-gravity. If gravity is an effect of energy, then anti-gravity is the way anti-matter experiences it as it travels backwards through time. Since energy experiences no time, and I suggest it is outside of time, there seems to be no way to conceive of anti, or negative, energy.

There's another problem with objects having potential energy which they may convert to kinetic energy as they fall. That is, what this potential energy is doing when it's not falling. Energy moves. We know that kinetic energy is the object either traveling or vibrating. What would be the movement of potential energy?

Suppose a rock floating in space is drawn to a black hole. If it's moving very little it has very little kinetic energy from travel. And it's cold out there, so the rock has very little heat, or vibrational energy. As the gravity of the black hole draws the rock to it, the rock accelerates to near light speed. By the time it reaches the perimeter of the black hole, it has kinetic energy which far exceeds its mass, proportionally, because it's going so fast.

Let's say it accelerates to 99.99% of light speed and its mass is 1 kg. We find its kinetic energy to be 6.28 x 10^{18} Joules, while the energy equivalency of its 1 kg of mass is only 8.99 x 10^{16} Joules. Using Einstein's formula for kinetic energy we find that the rock is 98.6% kinetic energy at this point. Its energy is 70 times its mass:

$$E_k = mc^2 \left(\frac{1}{\sqrt{1 - {v^2}/{c^2}}} - 1 \right)$$

where
  E_k = kinetic energy in Joules,
  m = mass in kg,
  c = the speed of light as 299,792,458 m/s,
  v = the speed of the object in m/s.

If falling is a process of converting potential energy into kinetic energy, then rocks floating through space are almost entirely potential energy. Our third rock from the sun would be also, and us. All objects we're familiar with seem to exist outside of black holes and other gravitational sources. Because gravitational bodies exist at all, any object is subject to this extreme acceleration.

In gravitational acceleration both objects participate. So if one were to give up energy to accelerate the other, then that one would also give up energy to accelerate the first. The result would be that each gives up some amount.

Could the energy to accelerate come from the other object in the attractive relationship? Could the black hole be giving up energy to the rock?

We seem to have one rock which is fully matter becoming mostly energy as it's accelerated by the black hole. It didn't have energy before the fall. The black hole is much larger than it. Might the black hole have given energy to the rock as it fell, similar to how a particle accelerator makes particles go faster in our laboratories by adding energy to them?

Imagine now, lots of these rocks far away from each other with nothing else around. A whole universe of rocks at rest in space. They, too, would accelerate to at least near light speed as they all were drawn together by gravity. How does each rock become mostly kinetic energy by the time they all collide? They can't get the energy from all the other rocks unless all the rocks are in energy form to begin with.

I suggest that potential energy is matter. Falling objects aren't getting heavier as they accelerate. They're not receiving energy from their attractor. They're just being converted into energy. Gravity converts matter into kinetic energy.

We know that we are matter, and that there are lots of objects in space which are also matter. We don't look up at the sky and see billions of objects made of only energy.

Gravity does not seem to be independent of its source. If it was to leave its source, then it would lose influence over it. Not only does a gravity wave not abandon its source, it seems to be a part of it. This is why gravity waves are able to carry with them the full energy of their objects. At large distances, the force of gravity is very small compared to the total energy of their object. But at a close enough distance the force of gravity is equal to the energy of the object. When you get close enough, an object does pull you with all of its might.

With our formula for gravitational force, we can see that it varies with distance. The closer we get to a gravitational body, the stronger the force is.

Newton's formula for gravity in combination with Einstein's formula for energy reveals the distance at which gravitational pull is the full energy of the object.

$F = GMm/r^2$ is expanded to $\quad E/r = GMm/r^2$.

Multiply both sides by r $\quad E = GMm/r$.

Replace E with mc^2 $\quad mc^2 = GMm/r$.

Divide both sides by m $\quad\quad c^2 = GM/r$.

Now we can divide both sides by c^2 and multiply both by r for $r = GM/c^2$.

The result is a distance of 7.4236 x 10^{-28} m for each kg of mass of the object. At this distance gravitational force is the full energy of the mass of the object.

If force cannot exceed the energy of the object, either we can't get closer to it or getting closer does not increase gravity.

It seems that the gravitational waves may be the mass of the object. If they are they must stay with their object. Otherwise, we have particle decay from this emission.

I believe this tells us what energy is doing when it's in particle form. We know that energy must continuously move at the speed of light. When matter moves, it does so in hops at light speed. While at rest in space the energy which the matter is composed of is traveling outward from its location in the form of gravity waves at the speed of light.

Radiant Energy

Radiation is emitted by a source and departs not to return again. It is emitted as a series of waves in a quantum of energy we call a photon. And it goes one direction.

Visible light, radio waves, X-rays, gamma rays, and microwaves are all radiation and leave their source. They are emitted as independent objects with no attachment, no need to return.

Force of Energy

Virtual photons which do the work of magnetism, appear surrounding their source and are said to be quickly re-absorbed. This is because forces are not energy but an effect of energy.

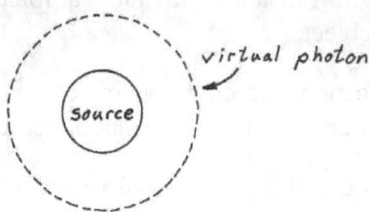

They may travel at light speed to reach their distance, their wavelength, but they remain encompassing their source. They can't rest because they're pure energy.

Gravity waves (gravitons) seem to behave similarly.

This wave, or area, of a graviton can be very large. We measure visible light waves in billionths of a meter. Radio waves can be measured in meters, kilometers or larger units. These waves can only be so long because they depart from their source.

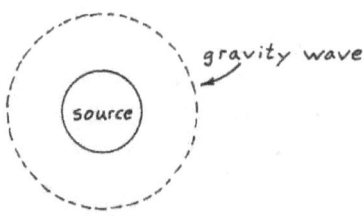

Gravity waves might possibly be measured in units as large as light years or more since they encompass the objects they affect all at once. They can be as large as the universe, but no larger than the distance which light could have traveled since the Big Bang. This is the limit of the reach of gravity.

LIGO

We see the speed and reach of gravity waves in data obtained by the Laser Interferometer Gravitational-Wave Observatory (LIGO). LIGO has experiments located in Livingston Parish, Louisiana and Hanford, Washington. Each is made up of an L-shaped arrangement of laser beam pathways. The beams contained in vacuum sealed tubes bounce back and forth between mirrors 2.5 miles away from each other. And very sensitive detectors determine whether there has been a shift of the beam in any direction.

Predictions are made for what a vibration from various gravitational events would look like and the system collects data for physicists to examine. Even with extreme sensitivity, they can only detect the strongest gravitational waves.

On September 14th, 2015, LIGO recorded a shift of their beams by one ten-thousandth of the width of a proton. The signal was seen at Washington first, then 7 milliseconds later (at light speed) in Louisiana. Its characteristics matched the description of a merger of two black holes of 29 and 36 solar masses. Since then, the detectors have continued finding similar events deep in space and equally far back in time.

Gravity Is A Force

I believe the success of the experiment at LIGO shows us that gravity is a force which affects objects in space and time, but does not affect space and time themselves. Our measurement system, the dimensions, is fixed.

Had space adjusted as gravity waves passed through the detectors, then the size of size would have changed, also. It's my understanding that space is distance itself. It's not an object. If size itself was to change, then everything would change and there would be nothing to compare the change to and reveal that there has been a change. If the length of a meter becomes smaller then bigger as a wave passes, then the detector would become smaller then bigger since its size is in meters. The laser beams, also sized in meters, would become smaller, then bigger by the same amount. Two waves at the same location in space would both shrink and grow exactly the same amount as space itself does at that location, and their alignment with each other or anything else around them would not change. The interferometer would detect no interference.

For the waves to misalign (for the device to detect a change), a mirror must move through space to change the distance the beam travels.

Because of this we know that gravity causes objects to move through space. While gravity is a force, I believe it works in a special way which does not cost energy. It couldn't, otherwise energy wouldn't be conserved. I believe gravity is energy waves coming into alignment.

While we can speak of gravitons as particles which are very small, their effect is as waves. For our purposes it's helpful to visualize them either as a bell curve or as an area.

With this visual we can imagine how they do their work. What we notice is that when the waves reach another object there is an overlap of the waves which coincides with the pulling force between them.

The relationship between a gravity wave and its emitting source is important to understanding the behavior of gravity.

One quality of these waves we can be certain of is that they will always encompass their object. They will never leave it. This is because they begin by surrounding it and travel at maximum speed (light speed) in all directions. The object, then, cannot leave those waves because it can't exceed light speed. It can't outrun its own gravity.

While waves should move according to where the object was when it was emitted, the object is still able to move. If we could see a moving particle and its waves, it may look like this.

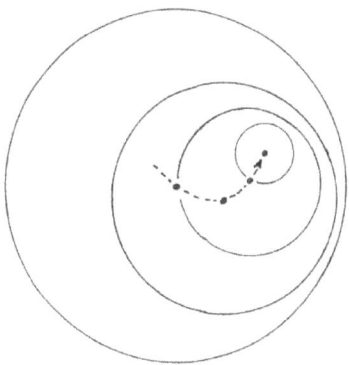

Each wave is centered on where the particle was in the past. And since the waves continue indefinitely through space, they represent the entire history of the particle.

Wave Alignment

Gravity seems to be the result of interactions between waves from each source. And I believe the interaction is the waves coming into alignment, similar to how other waves do. We can witness this with sound and light.

If you've ever heard two musical instruments coming into tune then you've likely noticed that two things happen. As the instruments are slightly out of tune while attempting to play the same pitch, or wavelength, the sounds produce an oscillation. The combined sound alternates between louder and quieter as slightly mismatched waves sometimes cancel and sometimes enhance each other.

loud   quiet   loud

Depending on how different their wavelengths are, the effect will be slow and wavy or quick as a vibration. But once in tune the sound is always loud as the two waves consistently enhance each other. In fact, they essentially merge into one bigger wave which is twice the height (or volume) of each of the individual waves.

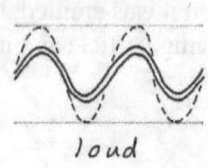

loud

This may not seem surprising because it's what harmony is all about. Harmony is enhancement.

Interestingly, it doesn't matter how close or far apart the musicians are, or where you are as you listen; they always sound loud together when they're in tune. It seems that waves of equal length should be just as likely to cancel each other as they are to enhance.

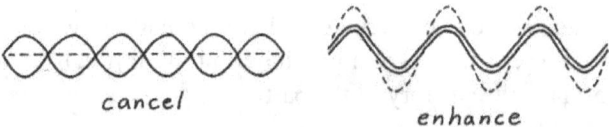

cancel   enhance

The two musicians coming into tune don't arrive at silence when they finally achieve waves of the same length which could cancel each other. It's always loud.

When ripples on calm water come from two different sources, each ripple travels outward from its origin at a fixed speed and wavelength forming rings around their source. This is the way gravity waves might appear if we could see them.

As each ring crosses over a ring from the other source there appears a cross-hatched pattern of enhancing and canceling. A speck of dust floating at one place would experience dramatic waves. At another location it may not experience

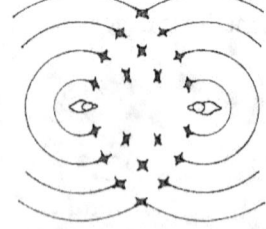

58

any waves as the crests and troughs align and cancel each other.

With this in mind, it seems the volume of a well tuned musical duet should depend on precisely where you're located in relation to the waves of each musician. Some notes should be louder and some quieter. The wavelengths of most musical notes can be measured in feet (middle C on a keyboard is about 4 feet), so it should matter where you sit in a theater, but it doesn't.

The waves always come into alignment with each other before they reach you.

Just as we don't find loud and quiet spots in a theater, we also don't find good and poor reception of radio or cell phone waves. As we drive on a road between two repeaters (antennas broadcasting the same signal) we don't lose reception regularly as their waves sometimes cancel and sometimes enhance each other. We find an equally strong signal as long as we're near an antenna.

Sound cancellation headphones are able to use out-of-phase waves to eliminate sound. Cancelation can occur in the short distance between your earphones and ear drums, but across a room it doesn't work.

Alignment

As the two sources emit waves and those waves cross paths they want to come into alignment. The waves, whether they're vibrations of air like sound, or electromagnetic waves like light or radio, travel at a fixed speed. They cannot speed up or slow down but their shape can adjust in order to align. We know they can adjust since we see it in gravitational color shifts.

Photons must move at light speed, but their waves adjust in response to gravity to create red and blue shifts.

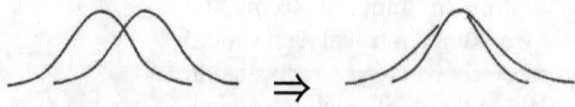

     Looking up at the lamp above me, my eye is connected to that light source by photons. Visible light wavelengths range from about 400 to 700 nanometers. (A nanometer is a billionth of a meter.) And a photon, the smallest piece of light, is about 10 million waves long. That means each photon coming out of that bulb and entering my eye is between 4 and 7 meters long, or between 13 and 23 feet. The light I'm receiving is being absorbed by my eye while it's still being emitted by the light bulb. It encompasses both the bulb and my eye at the same moment.

     This seems to be how magnetism works. The photons span two objects and push them apart or pull them together. I suggest that gravity works the same way. And the pulling together is based on waves wanting to align, or be in the same place, combined with the photons themselves encompassing both of the objects experiencing gravitational attraction.

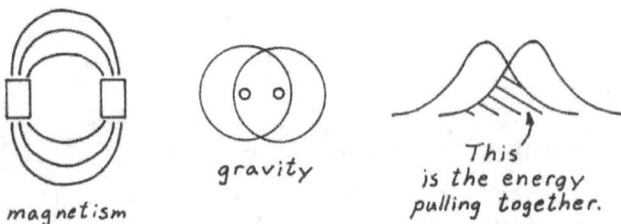

magnetism   gravity   This is the energy pulling together.

     The more the waves overlap, the more they pull. I believe what we see here is an application of a fundamental law of nature.

     When it comes to gravity, I believe this pulling together is done by the uncertainty principle. The more energy that is in a location, the smaller that location becomes. The shortening of wavelength due to greater energy is gravitational attraction. So, everything wants to come together, and we call it gravitational force. Only particle properties prevent them from getting too close. We call these properties electrical force.

It seems that gravitational attraction is just an effect of energy. It's not a special energy. It may be the energy of the object itself.

It appears that the principle which brought our universe into existence from a singularity, the Heisenberg uncertainty principle, also creates the effect we call gravity.

In the process, it also does something else. It reduces experience of distance and time.

Chapter 7 – Gravitational Dilation

We know that our experience of distance and time becomes less the stronger gravity we're exposed to. This is similar to the reduced experience we have from travel speed. (The faster we travel, the less distance and time we experience.)

While gravity is expressed as acceleration, which is speed over time, it can cause travel but it doesn't have to. Just because I'm not falling, or accelerating, into the earth doesn't change the dilation I receive from that potential travel.

I could imagine that I'm traveling. Perhaps falling into the earth but also accelerating away from it at the same rate. This way I stay where I'm at. The pull of gravity causes an object to have qualities which are the same as movement.

A traveling object is part energy and part matter. And its proportion of energy is equal to the distance and time dilation (reduction) which it experiences. This relationship is explained by movement theory.

Movement Theory

The theory of movement was created to fit together relativity and quantum behaviors, since on the surface they appear to be fundamentally incompatible.

According to quantum theory, space and time are not smooth environments. They are quantized. There is a minimum measure of distance and a minimum measure of time. This means we live in grid-like dimensions. And because of this I believe movement cannot occur in them.

Based on our quantization, in order to get from one grid space to another we must be hopping, or teleporting, because we can't be partially in one space and partially in another. We can't transition through space or time.

Meanwhile, relativity tells us that our physical laws are reliable in a smooth environment. We need a continuum. There must be a dimension which accommodates true movement, a movement dimension. And since this dimension cannot be quantized it can't be space or time.

I found that objects in our world move, or relocate, similar to chess pieces, by leaving the grid and coming back. The movement dimension is like the air above the chess board.

In this way space and time are seen as flat, fixed, and constant. They are just a measuring system. And since the grid doesn't allow for movement everything is either at rest in space and time, in the grid, or traveling outside of it at the speed of light.

I use hopping as a visual, but outside of space and time there are no particles. In this model, particles (matter) actually cease to exist very briefly as they relocate. This makes the speed of light the only speed there is. This absence due to hopping explains the loss of distance and time which traveling objects experience in proportion to their speed. And it accounts for every object seeming to be both a wave and a particle. In this model they must be both.

The wave is the moving aspect of an object and the particle is the presence of the wave in space and time. It gives the world we live in a flickering quality, like a film running at light speed. The frames in our world are a million trillion trillion trillion times faster than a movie since the smallest quantities of time are 5.391247×10^{-44} seconds.

Each particle of an object alternates between being energy which moves at light speed and being matter which is the same energy appearing to be at rest in space as a particle. In the grid of space and time, we could visualize it like this:

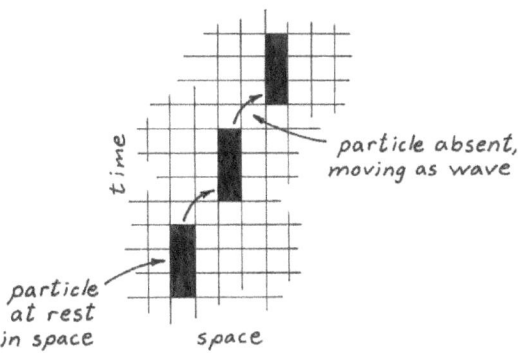

This explains why objects lose experience of distance and time as they travel. They're actually not in space or time during movement.

Furthermore, I found that if the particle aspect of an object was only the object's presence in space and time, I was able to keep the wave in place. It became the fundamental substance.

The wavefunction (the probabilities) never collapsed, as was proposed by the Copenhagen interpretation of quantum theory. It just provided us with an uncertainty of exactly where the particle would appear in space and time.

The Copenhagen interpretation was used by Werner Heisenberg and Niels Bohr in 1935 to explain how properties of particles could behave as if they are both precise and uncertain, how objects could be both a particle and a wave. They suggested that objects are uncertain, as a probability cloud, until they're observed. When observed, an object chooses specific properties (location, speed, and so on), the potential of it having other properties disappears.

The movement theory explanations provide a model which accommodates both the particle and wave aspect of objects. It is a continuum which allows our laws of physics to be reliable, and maintains the unpredictability of objects which quantum theory describes. With this in mind, if we consider gravity to be an interaction between waves, we can look at both the gravitons and the objects emitting them as waves.

Since everything in our world does have some degree of movement, even if it's vibration from heat, everything is in some portion energy and some portion matter. Everything can be considered to be alternating between being in and out of space and time. They can also be seen as particles which sometimes exist in space and time, and sometimes don't.

As I mentioned, when an object is in particle form its wave remains. The wave is just unable to travel while it's a particle. But it's still energy, so it's moving at light speed somehow. I believe that while an object is at rest in space as a particle, it's emitting waves as gravity. This leads to the conclusion that space and time dilation is equal to the proportion of energy and matter of an object due to travel. And since gravity is potential travel then maybe gravitational dilation is a result of the same energy and matter ratio.

Even though I'm not actually falling into the earth, maybe my ratio is affected just by being on earth. Maybe gravity causes me to be just a little more outside of space and time than I would be without it.

I believe we can learn how gravitational dilation occurs by looking at how dilation from travel occurs.

Dilation From Travel

Einstein's special theory of relativity told us that traveling objects lose time and distance according to their speed. The faster they go, the more dramatic their loss of time and distance is.

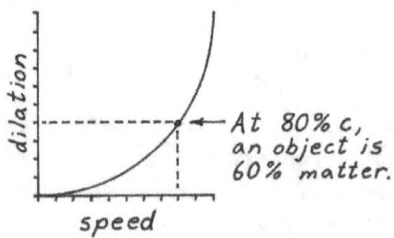

66

Movement theory tells us that objects are energy while they're relocating and are matter while they're at rest in space and time. This method of travel accounts for the loss of time and distance as a loss of experience; it's not an actual change in time or distance, as those are fixed.

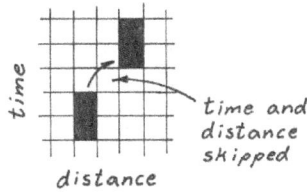

time

distance

time and distance skipped

Absolute relativity relies on this concept to explain how it is that space and time may be fixed, that there is absolute movement, yet we each have our own experience of it.

The formula which tells us how much time and distance is experienced was provided by Henrik Lorentz in 1895. It is

$$A = B \sqrt{1 - \frac{v^2}{c^2}}$$

where

A = experience of time or distance to the traveler,
B = (actual) time or distance to the bystander,
v = speed of travel,
c = speed of light.

This formula, then, tells us how much we are in space and time and out of it. It tells us of the hops. And because the hops represent being out of space and time (not in particle form) they also tell us of our proportion of matter to energy. More specifically it tells us how much of our whole being is matter because matter is experience.

$$\frac{matter}{matter + energy} = \sqrt{1 - \frac{v^2}{c^2}}$$

The portion of distance and time we experience is our portion matter (the amount we are in space and time) compared to our total matter and energy (the amount of actual space and time traveled).

67

If you could be perfectly still, have no energy, your portion of matter would be 1.0. One hundred percent of you would be matter. And your portion of experience would be 1.0. You would experience every moment. But if you were traveling 100 mph, your experience would only be 0.9999999999999889. You would only experience 99.99999999999889% of distance and time. That means you would be 00.00000000000111% energy. This is based on you traveling. I believe you also become energy just by being exposed to gravity because it causes you to lose experience too.

Dilation From Gravity

In order to understand how gravity produces effects similar to travel, we need to look at what gravity is. I believe it's an effect of energy, and as it pulls on objects it converts them into energy.

We saw how travel affects the proportion of matter to total matter and energy of an object. An object which is traveling is part energy and part matter. The part which is energy is referred to as kinetic energy, the energy of movement.

Gravity creates the effect of travel. I believe it, too, is altering the *matter/(matter + energy)* ratio of objects to accomplish this.

We know that objects which aren't traveling still have movement. Unless they have zero heat (0 Kelvin, -273.15 °C, or -459.67 °F) they have a type of vibration. And that is also kinetic energy.

Since gravity causes us to lose experience of distance and time like movement does, and gravity also causes movement, such as when we fall, then I suggest gravity causes us to vibrate when we're unable to fall. Our molecules move toward the earth, run into other molecules and bounce back. Our molecules fall all over each other in their attraction to earth. This is their movement. Gravity causes immobile objects to vibrate, though it's not the same vibration as heat, as we'll see later.

For our purposes we'll consider it as movement, or kinetic energy, which also includes heat vibrations.

This takes us back to Einstein's formula for kinetic energy:

$$E_k = mc^2 \left(\frac{1}{\sqrt{1 - v^2/c^2}} - 1 \right)$$

where

> E_k = kinetic energy in Joule-seconds,
> m = the mass of the object in kg,
> c = the speed of light as 299,792,458 m/s,
> v = the speed of the particles or molecules (or their average speed of vibration) in m/s.

And just as Lorentz's formula for dilation (loss of distance and time) due to travel tells us our portion experience, or *matter/(matter + energy)* ratio, this kinetic energy formula does also.

Since $\sqrt{1 - v^2/c^2}$ is experience while traveling, then

$$E_k = mc^2 \left(\frac{1}{\sqrt{1 - v^2/c^2}} - 1 \right)$$

can be expressed as

$$E_k = mc^2 \left(\frac{1}{experience} - 1 \right)$$

which is also

$$E_k = \frac{mc^2}{experience} - mc^2.$$

If we move the *mc²* to the left side

$$E_k + mc^2 = \frac{mc^2}{experience}$$

69

then arrange to solve for experience, we get

$$experience = \frac{mc^2}{E_k + mc^2}.$$

Since mc^2 is the energy equivalency of matter, we can call it E_m. This gives us

$$experience = \frac{E_m}{E_m + E_k}$$

which is *matter/(matter + energy)*.

(I use energy equivalency of matter rather than mass because each of our components need to be expressed in the same units in order to compare them proportionally. This could also be done using the mass equivalency of kinetic energy.)

We don't normally think of vibrations as movement because we can't see them. They're too small. But the effects which we notice on large objects actually are the result of what happens to their smallest parts. In this case, we're likely talking about molecules vibrating and it is real movement at that scale.

Traveling vs Vibrating

The difference between traveling and vibrating is whether the particles agree on which way to go. If they agree they accomplish travel.

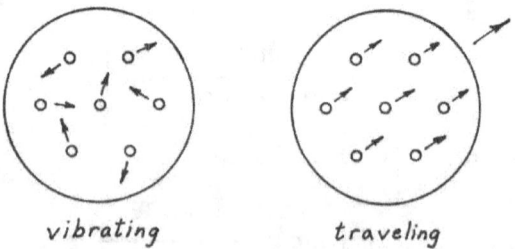

vibrating   traveling

This shows us how we can lose experience of time and distance due to gravity, just as we do from travel.

Our experience depends on what portion of us is in matter form. This is because experience comes from being in space and time, and matter is that presence in space and time.

Gravitational Heat

If gravity causes us to vibrate, making us part energy instead of entirely matter, then some of that vibration may translate into heat vibration.

On the surface of the earth, we experience a dilation of time and distance of 6.948×10^{-10} due to gravity from the earth. This makes our portion of energy at least 6.948×10^{-10}, or 00.0000006948%.

This amount of dilation is equivalent to traveling 11,200 m/s, or 25,000 mph. If all of this speed became heat vibration it would result in an intolerable temperature. The heat vibration of water in our bodies is only 682 m/s.

We can't see vibration from heat, and if we vibrate according to gravity we can't see that either. But since our body temperature is so low we know that any vibration from gravity is not the same as heat vibration. There's a way this is possible.

Everything in the world is both a wave and a particle. And these two parts of each object travel together. The wave moves in what seems like an orderly fashion according to the mechanical laws of our world. The particle, while with the wave, behaves erratically and gives us quantum probabilities. It's the particle aspect of an object which is unpredictable.

Heat vibration is real movement. We know this because it may be seen as what we call Brownian motion. It was discovered in 1827 by Robert Brown and explained by Einstein in 1905. You may see this for yourself by sprinkling a fine powder, such as pollen, in a glass of calm water. Viewed closely enough you'll see the pollen move around on the surface of the water as if it's alive. The pollen, even though it's much larger than the molecules of water, is being pushed around by those molecules as they move. The particle aspect of those molecules is moving and, likely, its wave is too.

As the particles encounter each other (what we would call bumping into each other) they transfer energy. This transferred energy is heat, which is low energy photons. It may also escape the water to energize the glass, air, or anything else around. Heat energy is communicated.

But energy which gravity generates is not communicated. It seems to be contained in the object. Perhaps it's expressed as vibrations within the wave of the object. In this way, the particle aspect of the water is moving erratically at a higher speed, but not encountering other molecules. For this reason, its energy is not communicated.

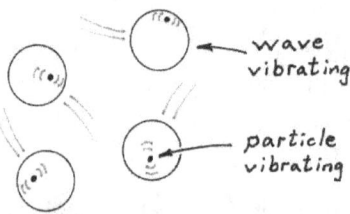

Since both heat and gravity seem to exist as an expression of energy, I suggest they may change into each other. Most likely, vibration from gravity may turn into heat. For now, we'll look closer at how the conversion from matter to energy occurs.

Chapter 8 – How Gravity Converts Matter Into Energy

I've explained how I believe matter, or the particle aspect of an object, is a phenomenon of waves, that it arises from energy. However, the objects still maintain a steady *matter/(matter + energy)* ratio which is not easily changed. If it could be, we wouldn't pay so much for gas and electricity.

We've learned from Einstein that as we try to make an object go faster it gains mass, or mass equivalency, and becomes harder to accelerate. This is because we add energy to it to make it go faster. Its mass, or amount of matter, remains the same. As we try to get it to light speed its mass equivalency advances toward infinity because energy has weight, too. Also, the only way to reach light speed is to become 100% energy. That means all mass must go away.

An increase in weight occurs if we push an object. This is because we're adding kinetic energy to it which adds to its total weight, or inertia. It takes more and more energy to accelerate it as it goes faster. Eventually it requires more than we can give it. Any amount of mass would require an infinite amount of energy to reach light speed.

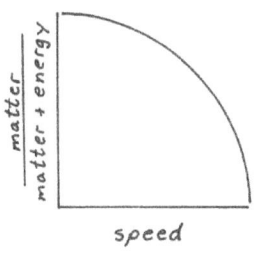

It seems that falling objects don't behave this way. An object falls with a constant rate of acceleration. The energy causing the force of gravity doesn't have to increase. It doesn't have a hard time making an object go faster just because it's already going fast. In fact, all objects accelerate at the same rate regardless of how massive they are.

It's true that increasing force is accommodated by the gravity formula, and the workings of gravity itself. Even if an object had started at a higher weight it would fall at the same rate. And if it was to gain weight on the way down, it will continue to accelerate at the same rate. But this can become a concern at high speeds. The great

73

amount of energy needed to accelerate at high speeds must come from somewhere, and the gravitational formula sets no limit for acceleration. (The formula places no limit on what it can accelerate or to what speed.) I don't believe objects become heavier as gravity accelerates them. And I think it's a mistake to believe that gravity adds energy to make them go faster.

We saw with the universe of rocks experiment that falling objects have no source adding energy to them. I suggest gravity does not contribute to or exchange energy with objects. It works by converting existing mass into energy.

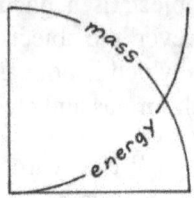

fall speed

Gravity is the effect of the energy waves of objects influencing each other. They are not giving them to each other. As part of their influence, they convert matter into energy so inertia remains the same. Since no energy is exchanged, no energy is consumed either. And we don't need an unlimited supply to keep objects accelerating. There is something else which could happen as objects accelerate and convert their matter into energy.

Particle Requirements

The different particles in our world seem to have a minimum mass requirement. To this mass, energy may be added in the form of heat or travel. But if the mass is decreased the particle may not be able to maintain its identity. For example, we know electrons to be 9.108×10^{-31} kg. If we convert that mass into energy, then we don't have an electron anymore.

This doesn't mean that when I throw a ball, its particles have to decay into photons. As I throw, I'm adding energy to the ball. And if I drop a ball, it also doesn't need to decay. The ball already has vibrational energy.

This may be converted into speed. If it is heat vibration which is converted, then the ball cools. But once travel speed exceeds the kinetic energy portion of an object, or its vibration speed, then travel speed must come from mass.

74

All speed may be converted into travel speed before any matter needs to be converted in a continually accelerating fall. Beyond a particle's vibration speed is where I suggest matter must be converted. That is the speed at which particles must decay.

Molecules are the pieces of matter we should expect to vibrate and carry heat energy. The particles which make them up (protons, neutrons, and electrons) have very fixed locations in molecules, so they're not free to move around. Molecules, while they make up the solid objects of our world, are less bound to each other so they may move around or vibrate. This is the reason why we can break rocks apart as easily as we do.

A molecule of water (H_2O) at room temperature has a vibration speed of about 667 m/s, or 1500 mph. Without adding energy to it, it should be able to accelerate to that rate before its molecules break apart. This applies to travel from acceleration and also vibration from an accelerating force, such as gravity.

Just as the dilation formula for travel tells us of the portion matter, so does the dilation formula for gravity.

Formula for Gravitational Dilation

The formula for determining dilation of time or distance caused by gravity is $1 - \sqrt{1 - 2GM/c^2 r}$.

This is similar to the travel dilation formula, so we may take it apart in the same way.

Since the formula above is dilation, or experience lost, then

$\sqrt{1 - 2GM/c^2 r}$ is our experience.

Experience while traveling is $\sqrt{1 - v^2/c^2}$.

This tells us that $2GM/c^2 r$ is the v^2/c^2 part of it.

v^2/c^2 is speed as a portion of light speed, and it is squared. So we can remove the speed of light and see that $2GM/r$ is v^2.

This means the speed which gravitational dilation is based on is $\sqrt{2GM/r}$. Notice that this is the escape speed formula.

This is more easily understood in the form of $\sqrt{GM/(1/2\,r)}$. It tells us the average speed caused by gravity while falling a distance of r. This speed is the speed reached in a fall of a distance of $1/2\,r$.

The gravitational dilation formula tells us the dilation we would be subjected to for our average speed of falling to the center of gravity.

I suggest dilation occurs because this is our actual speed. But since we're not falling, we're vibrating at this speed. This vibration is energy and it comes by converting mass. I suggest that gravity provides distance and time dilation by converting matter into energy. This means that while all the energy in the universe isn't enough to push a single particle up to light speed, gravity can pull any amount of matter up to light speed, even the whole universe as we'll see later.

When we evaluate conditions around a black hole we can find significant vibration speed among objects. If an object could be at rest at an event horizon it would experience 100% dilation from the gravitational pull. This means that the vibration speed is already 100% of light speed. It doesn't need to travel at all. An object at an event horizon converts into energy just by being there.

I've suggested that gravity converts matter into energy and shown how to determine the conditions which result in it. Now let's see how the conversion takes place.

76

How Gravity Waves Convert Matter Into Energy

It is commonly agreed that each particle continuously emits waves of energy. And those waves result in objects pulling together.

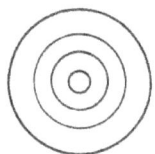

I suggest that if those waves are similar in alignment with waves of another object, then they pull each other to match. Furthermore, any overlap of the waves creates some amount of pull.

The larger waves are closer to being in alignment than the small ones. At a distance these are coming into alignment. At closer range, the small ones pull into alignment, also.

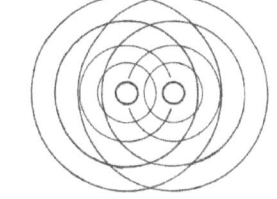

As on a pond, they adjust to fit each other. While troughs will fit with peaks to cancel, peaks are where the highest energy is and they seek to align with other peaks.

As waves pull each other together, they attempt to be smaller. Wavelength tries to shorten because the compromise between two slightly out of phase waves, even if they're the same size, is to narrow their peak.

This appears to be the mechanics of the Heisenberg uncertainty principle. The greater the energy, the smaller the area it occupies.

While molecules of air and water may bounce off of each other because they can't exist too closely, their waves want to share the same location. This is the reason why a singularity of energy is content. But gravity waves are still a part of their object. They are caused to be a shorter wavelength and wavelength is inversely proportional to energy. This relationship explains how changing the shape of its wave can change the energy of the object.

Wavelength/Energy Relationship

The relationship between frequency of waves and length of waves is through the speed they travel at. So for light $f = c/\lambda$ in which f is the frequency, c is the speed of light and λ is wavelength.

Putting this concept together with Planck's formula of $E = hf$, we can see that $E = hc/\lambda$. As wavelength shortens, energy goes up. The pulling together of gravity waves has increased the energy of each object by gravity waves shortening.

But total mass and energy of the objects must be conserved. Energy can't just be added to the objects to increase their total energy and therefore increase their total mass and energy combined. The energy must come from somewhere. So, the matter quantity must be reduced.

Waves aligning with waves causes the energy of objects to increase. Conservation requires that mass be given up. We can't have more energy without losing mass. And the higher ratio of energy must be expressed as movement, either travel or vibration.

This tells us that when two waves encounter each other, they shorten their length. With a shorter wavelength they have more energy. And along with that is less mass. Gravity causes objects to convert matter into energy and express it either by falling or vibrating.

Wavelength is allowed to shorten dramatically for an object because while momentum occurs linearly, the matter-energy equivalency is exponential. If we reduce a wavelength to half, the object's momentum doubles. But it doesn't have to give up half of its mass.

Suppose we had a particle of 10^{-30} kg mass moving at 53,000,000 m/s. It has a wavelength of 0.002 nanometers. If we reduce its wavelength to half, which is 0.001 nanometers, without giving any energy to it, it has doubled its energy but only needs to give up about 5% of its mass. We can see from this illustration how changing wavelength and changing the *matter/(matter + energy)* ratio are related.

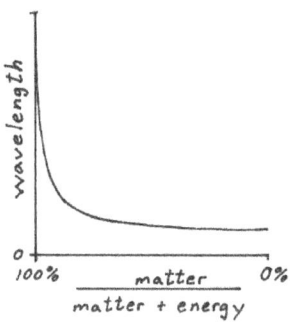

While we're discussing the behaviors of waves, let's look at another trick they're capable of.

Chapter 9 – Escape from the Inescapable

Black holes emit gravity. Even from their center where nothing can escape from, somehow gravity does. We know that gravity is not immune to the attraction of a black hole because gravity is the attraction. It is waves interacting with waves. And since gravity is an effect of energy, the energy of the black holes must get out. Black holes, even though they're inescapable, continue to interact with the rest of the world. It seems they should disappear from existence, swallowing all of their own waves, yet their influence continues to tell us that they are not only there but fully interacting with their surroundings.

Remember that our universe may have begun as an inescapable environment, a singularity of infinite force. And yet it escaped too.

We need to consider how energy is emitted from objects to understand how this is possible.

In order to escape from the inescapable, quantum uncertainty is invoked. Since it is not a force it doesn't have to overcome or work against the gravity which holds black holes together. Objects within a singularity or other confined space, such as a crowded atomic nucleus, cease to exist within that space and proceed to exist outside of it. Energy in the form of gravity waves does the same.

Consider what a confined space is. A singularity of the whole universe, a black hole, and an atomic nucleus are confined. Each is a quantity of matter and energy within a limited region. The quantities and sizes are all different, and so is their density, as well as the ratio of *matter/(matter + energy)*, velocity, and other properties.

The only factors involved in using quantum uncertainty to escape are momentum (the amount of energy and matter, with its average speed) and the size of the space.

The key to gravity waves getting out is that while they are

emitted in an inescapable environment, the large waves don't come into existence (into space and time, or into influence) until they're outside of the inescapable area. According to movement theory, if their wavelength is larger than the perimeter, then they pass right through it before coming into space and time. They are not traveling to get out. That would not be possible. These gravity waves don't have to escape from a black hole because they are larger than it, so they're already outside of it.

This is a very important concept in energy. While at first it seems that all energy is the same, only wavelength is different, wavelength determines the application of it.

Only very short wavelengths can be used to create the effect we call the strong nuclear force. Wavelengths coinciding with electron orbits may be used for chemical interactions. Longer waves must be used to create electrical flow, such as in communications. Gravitational waves are often too big to even have communication value. They can be so big that they are of no practical use except that they penetrate and escape any environment to give us gravitational force and direct our cosmic bodies.

This leads us to consider what other waves may be able to get out.

Communicating Out of a Black Hole

If a black hole can contain any sort of matter or energy, like stars, planets, and galaxies, and gravitational waves can escape from that environment, then other waves can, too. This would mean one could communicate out of a black hole.

The way gravity waves are able to escape a black hole is simply by being bigger than the inescapable horizon. This should not be a trick exclusive to gravity waves. Communication (radio) waves could apply the same principle.

Certainly any size wave which encounters a black hole could enter it. If a person was inside a black hole with a radio receiver, he could receive any signal because they can all come in. And by generating his own electromagnetic waves which are larger than the horizon which contains him, he could communicate back out. Radio communications out of a black hole should be possible as long as the communications are of a wavelength greater than the size of the inescapability.

While gravity waves generally operate at a larger scale than communication waves, we should expect their sizes to overlap. Certainly there are black holes smaller than our communication waves. We currently communicate with radio waves as large as several meters long. If the earth was a black hole it would be less than 3/4 of an inch in diameter. Radio waves, and even some microwaves, should be able to escape.

Matter Can Escape

Matter is always a wave according to movement theory. The particle is just the wave's presence in space and time. I believe this is how matter escapes an atomic nucleus, and I believe it should also be able to escape a black hole.

To understand how this can happen, we should consider the situation to be similar to the decay of a radioactive atomic nucleus. The method of escape for a piece of matter is the same as it is for energy.

Because a neutron can develop a wavelength, or probable location, which is larger than the nucleus it's confined to, it can escape even as matter. This escape of matter is the source of much radioactive radiation. Neutrons escape as matter from dense nuclei, then quickly decay into pure energy, very high frequency (dangerous) radiation.

Just as gravity waves do escape, and communication waves should escape a black hole by being larger than its perimeter, matter waves should also.

All matter is both a particle and a wave. According to movement theory the wave is the underlying structure, the probability, and is always there. The particle, or matter aspect of an object, only appears in space and time periodically. It is the wave's appearance in our world. Movement is the object coming into and going out of space and time as it travels.

In order for matter to get out of a black hole, its particle doesn't need to. Only its wave needs to. The wavelength of that particle must be greater than the inescapable size.

To have a large wavelength, momentum must be low. Since momentum is mass times velocity, one of these needs to be reduced dramatically. In order to get matter out, mass must be kept, so speed must be reduced.

Some amount of speed is caused by gravitational force, whether it be travel or vibration speed. So in order to eliminate speed, not only does the object need to stop traveling, gravitational force must be reduced so that it can stop vibrating too. There's only one place within a black hole, or any gravitational body, where gravity can be zero. That's the center.

Regardless of the amount of mass an object has, its momentum can be reduced by decreasing speed. This means eliminating movement. Size is inversely proportional to momentum. Since momentum is mass times velocity, no speed equals no momentum. Of course no momentum is impractical, but a very low momentum may be possible.

Any object which can position itself at the center of a black hole and cease nearly all of its movement will increase its wavelength, the area where it may be found. It may then appear in particle form anywhere in that region, possibly outside of its confinement.

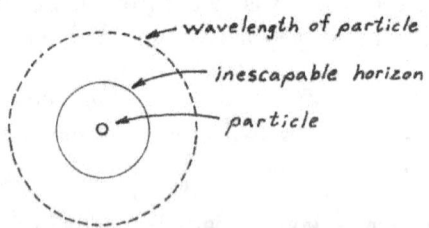

By achieving a near 100% experience, matter may escape a black hole.

A problem with escaping in this way is that the particle has no speed and it's likely near the black hole it just escaped from. Without quickly being given travel energy to leave the area, it will be drawn back in.

There's another inhibiting factor and that is the rest of the mass of the universe. If the black hole is anywhere other than the center of the universe, then any object inside it still has dilation from the mass around it. This means the location where a particle of matter can escape from any confinement is very special.

This leads us to the cancelation of gravity. If gravity becomes zero at the center of a body, then is gravity really gone or is it just balanced?

Chapter 10 – Canceling Gravity

Gravity has been said to be equivalent to acceleration. However, we know that it's a force which causes acceleration and is not just acceleration. First we'll look at how we know that it's more. Then we'll consider what acts against it.

More Than Acceleration

The earth accelerates all objects near its surface at 9.79 m/s^2. It also causes those objects to lose experience of distance and time in the amount of 6.95 x 10^{-10}. But the dilation isn't caused by acceleration alone. It also involves distance. We can see this by recreating our gravitational acceleration in different ways.

The earth has 5.97 x 10^{24} kg of mass and a radius of 6,378,000 meters. This mass and size gives us our acceleration, the pull of gravity which we feel. But we can also create this same acceleration with a much smaller mass.

A mass that is 1.47 x 10^{11} kg with a 1 meter radius would also provide 9.79 m/s^2 of acceleration. And it would only dilate by 2.18 x 10^{-16}. A much greater mass can result in the same acceleration, too. 3.09 x 10^{42} kg with a radius of 4.59 x 10^{15} meters would do it. It would dilate any object at that distance by 100%. Those objects would experience no time or distance by just sitting there. In fact, that mass would be a black hole. If it was a planet we could sit on its surface and feel the same gravitational pull we feel now. While it wouldn't pull us into its center any more than the earth pulls us to its center, it would convert our mass fully into energy.

So gravity is not the same as acceleration. Acceleration is just one of the consequences of gravity. When we look at what may cancel gravity we have to consider whether both acceleration and dilation are being eliminated.

There are ways of canceling only acceleration and not dilation, and they can mislead us into believing that gravity is canceled, or its effects eliminated.

Centrifugal Force Doesn't Cancel Gravity

A body in orbit is balanced between two opposing forces also. Gravity pulls it toward its attractor and an equal amount of centrifugal force pulls it away.

We know that centrifugal force doesn't cancel the dilation caused by gravity because our GPS satellites still experience it and have to adjust for it. And they are in balanced orbits. This tells us that gravity is not just acceleration. It is not the same as every other accelerating force.

While centrifugal force causes acceleration, it's not a direct force. It's a derived force.

There is nothing pulling the satellite away from the attractor. Centrifugal force is a geometric interpretation of inertia. And inertia is not acting against gravity either. It's acting perpendicular to it. This creates zero opposition to gravity.

Gravity is not pulling the satellite down. It's holding it down just as it's holding me down on earth right now.

Satellites are often described as perpetually falling. This is because the mathematics can be viewed as equivalent. But they aren't getting any closer to the earth. Gravity curves their path, which is determined by inertia. It holds satellites against escape by acting perpendicular to it. If an object can't move toward its attractor it must vibrate.

I believe that if I threw a ball upward the ball would be dilated by gravity on its way up, just as it is while it's at rest on the ground. It would vibrate. At the top of its travel, it would vibrate also. And as it began falling its vibration would turn into downward travel speed. Gravitational dilation becomes travel dilation as an object moves

88

toward its attractor.

Gravity Cancels Gravity

When multiple gravity sources pull an object in different directions, those forces cancel. The same would occur while inside a gravitational source, surrounded by its mass. We saw this when we imagined traveling into the earth.

I suggest that not only is there a cancellation of pull, but a cancellation of gravity itself. And this reinforces to us that gravity is not just acceleration, and is also not the waves. When gravity cancels, I believe it is actually gone.

It was once predicted that the center of the earth is 2 1/2 years younger than the surface of it. This was based on the idea that there should be more gravity at its center. If gravity was present simply because of the presence of mass, then there would still be time and distance dilation resulting from it even while its pull cancels. Time dilation, the loss of time due to the presence of gravity, is what would cause the center of the earth to be younger than its surface.

The pull we feel from a massive body such as the earth decreases according to the inverse square law as our distance from the center of the source (the earth) increases. It's not our distance from the earth we calculate, it's our distance from the earth's center.

But that doesn't work once we go into the object we're pulled toward. If we dug a hole in the earth to get closer to its center, gravity would not increase. It would decrease. This is because part of the object would be behind us pulling us in the other direction. As we reach the center of the earth, gravity becomes zero. At the center, we have an equal amount of mass on all sides of us, so we wouldn't be pulled in any direction. We would be weightless.

Here on the surface of the earth, we experience 6.95×10^{-10} dilation due to gravity. Could we still experience this dilation, or even more of it, while also being weightless at the earth's center?

I don't believe the center of the earth is 2 1/2 years younger, but 3000 years older, because I believe the dilation goes away, too. (3000 years is the dilation the surface of the earth has experienced since it formed.) It goes away because it's caused by the force and the force cancels.

Consider that we ARE surrounded by gravitating mass right now, a lot of it, the whole universe. As I stand here on earth, I feel none of it. I don't feel pulled in any direction except down to the earth.

There is a pull that the sun and moon have which causes ocean tides to rise, and we get effects from other planets too. However, we have no direct evidence that the gravity of the rest of the universe exists at all. We count and measure the stars and galaxies and calculate that we're surrounded by over a hundred thousand trillion trillion trillion trillion kilograms worth of stars, planets, black holes, dust, and other objects. They all have gravitational force. That's what holds them together and moving around each other. And their gravity can reach us, too.

Since we don't have any effect from it, it must cancel. And the more completely it cancels the more centered we must be in it.

The pull of gravity can be countered by centrifugal, or outward, force if we're traveling around a source of gravity, such as the sun or center of our galaxy, or going around the universe. But gravity has two effects on us. It pulls us and it takes away our experience of distance and time. Centrifugal force could counter the pull, but it couldn't counter the loss of experience. In fact, if we were orbiting the center of the universe, or any other gravitating body, we would have speed in addition to the pull of gravity contributing to our loss of distance and time. Orbiting to balance the pull of gravity compounds the problem of losing experience. This loss cannot be countered because there is nothing which increases experience. We need to look at how much dilation we experience, then.

In our daily lives we may not know if we're losing distance and time experience since our whole experience of the world adjusts together. If we only experienced 1/2 of every minute, we wouldn't know. Everything around us would also experience 1/2 of every minute, including our clocks. If we were very dilated (reduced in distance and time) we might live our lives normally. But a bystander, perhaps someone at rest in space looking at us through a powerful telescope, would see us as living in slow motion.

There is a way to tell if we're living a dilated, or slow, existence and that is by measuring relative dilation. Since dilation occurs exponentially as gravity or speed increases, we can use differences in gravity or speed to tell us where we are on this exponential curve.

At low total speed or gravity we get a small amount of dilation from large speed or gravity differences. At high total speed or gravity we get a large amount of dilation from a small amount of speed or gravity difference.

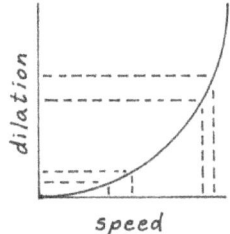

We already have experimental data of the dilation differences we experience.

Hafele And Keating Experiment

In 1971, Joseph Hafele and Richard Keating flew a cesium clock around the world both east and west to study dilation from gravity and speed. The results show that the earth rotates each day, but does not orbit the sun. It also shows that we don't have enough dilation for us to be anywhere but near the gravitational center of the universe. We are at, or near, the bottom of the dilation curve.

I estimated their data to be:

Eastbound
- altitude of 30,000 feet (9146 meters),
- flight speed of 570 mph (917 km/hr),
- earth's rotational speed according to latitude of 1344 km/hr,
- total flight distance of 24,860 miles (40,000 km),
- total flight time of 43.61 hours.

Westbound
- altitude of 35,000 feet (10,670 meters),

- flight speed of 510 mph (821 km/hr),
- earth's rotational speed according to latitude of 1344 km/hr,
- total flight distance of 23,617 miles (38,000 km),
- total flight time of 46.31 hours.

They found that the flight clock lost 59 nanoseconds (billionths of a second) more than the earth clock going east and lost 273 nanoseconds less going west.

The results show that the earth does rotate, because by going both directions they were able to achieve notably different results. Most significantly, the flight clock lost more time compared to the earth clock by going one way and lost less time (showing a gain) going the other way. The fact that it can change from losing more than the earth clock to losing less by changing direction means that the speed changes of the plane are significant compared to the speed and gravity of the earth. At a few miles altitude and several hundred miles per hour, they are comparable to the total gravity and speed we're exposed to. Had total gravity or speed been greater, meaning we're exposed to a lot of it, all flights would result in less loss of time than the earth clock.

For example, had the earth been orbiting the sun (at 107,000 km/hr) they should have found a loss of 29,033 nanoseconds less going east, and a loss of 7332 nanoseconds less going west. This doesn't mean that the earth isn't moving through space, but that it certainly isn't moving that fast.

Just because we're not going east or west fast enough to affect the results doesn't mean we're not traveling north or south, or experiencing extra vibration due to gravity. But it means it's not significant. If it was making these movements, they would shift our results toward a greater loss as they would push us further up the curve. Being higher up the curve would make any change in speed or gravity more dramatic. It would make clocks flying either direction lose less time than the earth clock. If we were located 00.01% from the center of gravity of the universe, the results would have shown the eastbound flight clock to lose 467 ns less than the earth clock, and the westbound flight clock to lose 923 ns less than the earth clock.

While this doesn't tell us for certain that we're experiencing no other speed or gravitational dilation it does mean that if we were moving a few thousand miles per hour in any direction, or were located anywhere other than the center of the universe, all results would be notably more dramatic.

Temperature

Another thing we can see from this experiment is that heat vibration may affect dilation.

The data from Hafele and Keating could allow for the vibration speed of the clock's atomic material, cesium-133. The clock measures time by the decay of cesium, so it measures the time those atoms experience. At room temperature they should have a heat vibration of 192 m/s (430 mph). This much speed combined with gravity and travel speed would not dramatically change the results of the experiment. It would only push the results slightly up the curve.

Gravity of the Universe

We've seen that acceleration cannot cancel gravity and looked at data from an experiment proving the effects of gravity. Now let's reconsider the gravity of the universe and whether it's being canceled.

We're surrounded by the universe. Whether we're at its center or not, there certainly is plenty on all sides of us. So let's find out how much dilation all this matter should give us. First we'll assume that we're at its center.

The mass of the universe is estimated to be 1.73×10^{53} kg. And its radius is believed to be 1.28×10^{26} meters (13.8 billion light years). If we are located at the center of it and its mass is distributed evenly, as it appears to be, the average distance to this mass would be 1.059×10^{26} meters.

I calculated the volume of the universe, then found the radius of a sphere of half that volume:

$$4/3 \, \pi \, r^3 = \text{volume so } \sqrt[3]{\frac{3 \cdot volume}{4\pi}} = radius.$$

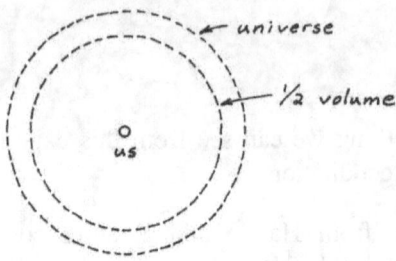

The gravitational dilation formula tells us we should experience a dilation of 100% from it. Now let's suppose we're at the edge. We can calculate this the way we do the gravity on the surface of a planet. If we consider the universe as a large body with its size and mass, we find that we're still 100% dilated by its gravity.

Obviously, we do experience distance and time. And we're not way outside of the universe looking in; we're well within it.

We don't seem to experience any dilation from the mass of the universe. I believe this tells us two things.

1. Gravity cancels gravity. It doesn't just counteract it as an opposing force.

2. We are located at or very near the center of the universe. This is why it cancels so completely.

How Gravity Cancels

If one friend pulled on one of your arms and another friend pulled on the other, you wouldn't go either direction. But you would feel pull because each is only pulling on part of you. You would feel the separation of your parts.

If we pull on a fundamental particle both ways it can't go either way, or feel a pull. It can't be pulled apart. Pulling every individual particle has no effect at all. You only feel it because your

94

friends pulled on your arms not the whole you. Had they been able to pull on each of your particles equally you would not have felt it.

Gravity occurs at the most fundamental level. Every particle pulls on every other particle. We just describe it for convenience as large objects pulling on large objects.

While we can imagine a planet, or star, as having its own large powerful gravitational waves, they are actually a composite of those of its particles.

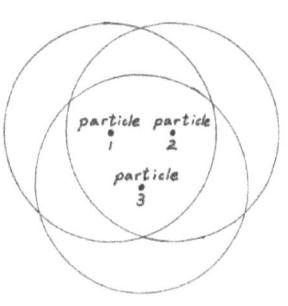

At the center of gravity particles are pulled equally in all directions. The result is that they don't go anywhere. They're not made to fall or vibrate by this balanced pulling. They don't even change size.

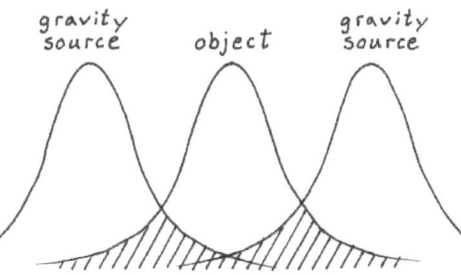

With equal energy on all sides, what's the object to do?

Gravity Is Not An Object

If gravity was an object of its own, then certainly anything surrounded by matter or energy would experience a lot of it. Instead because it is able to cancel, we know that gravity is not the waves, but the effect of the waves.

The reason why we change our ratio of *matter/(matter + energy)* when we're pulled by gravity is because the pull causes our particles to move. It causes either travel or vibration.

It's the mobility of particles which allows this. If solid objects were made of particles with no space between them, then we wouldn't have vibrational energy, only travel. The fact that matter exists as waves allows for matter to express gravitational dilation.

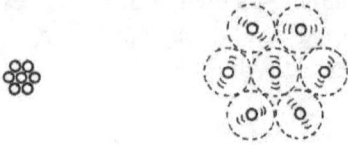

We've seen further evidence that gravity is not an object, but an effect of energy. And because of this it's able to cancel. This cancelation of gravity tells us that we're at the center of the universe. And if the universe rotates or expands, our lack of travel speed also tells us that we're at the center of it.

Now let's look at that rotation and see how it's affected by gravity.

Chapter 11 - Orbits

Earlier we saw the need for every energetic system to rotate. Planets and stars spin on axes. Solar systems and galaxies rotate. And the whole universe should be expected to turn as well. So, now we'll discuss extreme gravitational systems from the perspective of orbits. The reason is because orbits are very common and they're also useful in preventing collapse from gravity.

Orbiting a (Singularity) Black Hole

We often define a black hole as a mysterious sphere, an event horizon which nothing can escape from. So we can only speculate about its contents.

In order to explore orbits around an extremely dense object I want to begin by imagining that there exists a quantity of matter which is very compact. We can consider it to be well within the boundary of its Schwarzschild radius. This way we can consider it as a singularity, but it's really just small enough that we don't have to worry about going inside it where gravity becomes less.

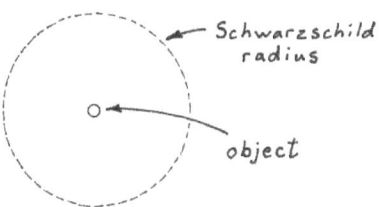

As with any object of significant mass, other objects of lesser mass may orbit it. I'll use Newton's formula for gravitational force, GMm/r^2, to determine how strong gravity is in various orbits. And I'll use mv^2/r for centrifugal force with it to tell us how fast we need to travel in order to orbit at that distance. To establish an orbit, centrifugal force must equal gravitational force. We can combine these two formulas to tell us that the orbit speeds are the same for any object at the same distance.

Since their force must be equal to establish an orbit, then $GMm/r^2 = mv^2/r$. We can cancel the mass of the orbiting object (m) since it will be the same amount on both sides. This tells us that orbit speed is dependent only on the mass of the object being orbited, and distance from it. The mass of the satellite doesn't matter. We can also cancel one of the appearances of distance (r) since that will also be the same:

$$\frac{GM}{r} = v^2$$

To solve for speed, we can arrange it as $v = \sqrt{GM/r}$.

With this formula we can determine the necessary orbit speed given any particular mass and distance. In order to express our distances as portions of the Schwarzschild radius, we can insert that formula in place of r along with a factor of p_S for our portion of the Schwarzschild radius:

$$v = \sqrt{\frac{GM}{p_S \frac{2GM}{c^2}}}$$

Using portions of the Schwarzschild radius, the mass or size of our black hole no longer matters. The gravitational constant (G) and mass variable (M) which determined its size will cancel:

$$v = \sqrt{\frac{c^2}{2p_S}}$$

This gives us the following orbits as portions of the Schwarzschild radius. They may be applied to any black hole which has a very compact mass.

Radius R_S	Orbit Speed	Radius R_S	Orbit Speed
0.50	1.000c	1.75	0.534c
0.75	0.816c	2.00	0.500c
1.00	0.706c	2.25	0.471c
1.25	0.632c	2.50	0.447c
1.50	0.577c	2.75	0.426c
		3.00	0.408c

(I've divided v by the speed of light, c, in order to express these speeds as portions of light speed.)

We can see that at 0.5 of the Schwarzschild radius an object would need to travel at light speed in order to stay in orbit. At twice the Schwarzschild radius, it would have to travel 0.5 of light speed.

These speeds tell us that there may be realistic orbits both outside and inside of a black hole perimeter. Not considering other factors, it appears that a piece of matter may orbit at, or inside, this radius. And the real limit for orbiting this type of black hole is 0.5 of the Schwarzschild radius.

So far we've only considered gravitational pull and speed of orbit. Gravitational dilation is also a factor in what can exist in these extreme conditions. This is because we're looking at high speeds and high gravity. Both affect the *matter/(matter + energy)* content of an object.

With Dilation

I've expanded our table to include dilation from speed and also from gravity to see how much each of these factors converts matter into energy. I've used

$$1 - \sqrt{1 - v^2/c^2}$$

for dilation due to speed and

$$1 - \sqrt{1 - \frac{2Gm}{c^2 r}}$$

for dilation due to gravity.

In order to standardize gravitational dilation for our black hole radii, I placed the Schwarzschild formula in place of r and added a portion variable of p_s:

$$dilation = 1 - \sqrt{1 - \frac{2Gm}{c^2 p_s \cdot \frac{2Gm}{c^2}}}$$

This simplifies to

$$dilation = 1 - \sqrt{1 - {}^1\!/_{p_s}}.$$

With these formulas for dilation due to travel and gravity we find the following:

Radius R_S	Orbit Speed	Dilation From Speed	Dilation From Gravity
0.50	1.000c	1.000	1.0000
0.75	0.816c	0.422	1.0000
1.00	0.706c	0.292	1.0000
1.25	0.632c	0.225	0.553
1.50	0.577c	0.183	0.423
1.75	0.534c	0.155	0.345
2.00	0.500c	0.134	0.293
2.25	0.471c	0.118	0.255
2.50	0.447c	0.105	0.225
2.75	0.426c	0.095	0.202
3.00	0.408c	0.087	0.184

When we look at dilation from gravity, it appears that matter cannot orbit inside an event horizon. This is because being inside that radius would cause it to fully convert into energy just by exposure to the force of gravity. The only way to avoid this is by falling toward its center.

Combining Dilations

And now, with two different dilations for an orbiting body we need to find a way to combine them. I believe each object can only have one experience, or *matter/(matter + energy)* ratio, based on one total dilation. And I don't believe these two dilations should be simply added together.

Remember I suggested that being under the influence of gravity causes dilation by causing actual speed, not just the effect of speed. And this speed expresses itself as vibration. So, while we're orbiting a black hole experiencing travel and vibration our total speed goes up. Because vibrations go all different directions and travel only goes one, we can't just add these speeds. But there is a way to combine them.

We had previously found our vibration speed to be determined by

$$\sqrt{\frac{2Gm}{r}} = vibration\ speed\ .$$

(This is the same as the escape speed or maximum speed attainable without escaping.)

Now we can use this formula to find speeds at various portions of the black hole radius by replacing the r with p_S for portion, along with the Schwarzschild formula:

$$v_v = \sqrt{\frac{2Gm}{p_S\ \frac{2Gm}{c^2}}}$$

And it simplifies to

101

$$v_v = \sqrt{\frac{1}{p_s / c^2}}$$

or

$$v_v = \sqrt{\frac{c^2}{p_s}}.$$

It may be used to determine our rate of vibration at all radii as portions of the Schwarzschild radius.

Once we have vibration speed it can be combined with travel speed, but we have to consider the direction of the vibrations. They may go in any direction. We can expect their directions to be random or based on very complex interactions making them unpredictable. So we must assume they go all directions equally.

If we average together all possible vibration directions from 0° to 180° relative to travel direction and weight them according to the relative circumference at that angle, we have a formula for combined speed which looks like this:

$$\frac{\sum_{\theta=0}^{\theta=180} 2\pi \frac{v_v}{c} \sin \sqrt{\left(\frac{v_v}{c}\cos\theta + \frac{v_t}{c}\right)^2 + \left(\frac{v_v}{c}\sin\theta\right)^2}}{\sum_{\theta=0}^{\theta=180} 2\pi \frac{v_v}{c}\sin\theta}$$

where

v_t = travel speed,
v_v = vibration speed.

The function $\sum_{\theta=0}^{\theta=180}$ means add up all solutions to what follows for every integer angle (θ) from 0° to 180°. The process may be shortened by using 5° or 10° intervals.

There is a shorter method to accomplish this combination. The results of this vector formula can be very closely approximated using the Pythagorean theorem. It only requires a downward adjustment of about 5.5% for the result.

We can see the relationship of these speeds in the common form of

$$\left(combined\ speed \middle/ 0.945 \right)^2 = v_t^2 + v_v^2 .$$

And we can arrange this for use as

$$combined\ speed = 0.945\sqrt{v_t^2 + v_v^2} .$$

A Note About My Hafele and Keating Data

I arrived at these flight data differently than I had in my book *Absolute Relativity*. For these, I determined the speed responsible for gravitational dilation and combined that with travel speed. Then I found dilation based on the combined speed. This most affected the eastbound flight. My previous method was to determine dilation from travel and dilation from gravity, then add the two results together. I suspect that is common practice since they are currently believed to be separate phenomena.

It is interesting that combining the effects of gravity and travel in this new way gave me a greater loss of time for the eastbound flight than the previous method did.

This coincides with Hafele and Keating under-predicting their results for that same trip. (They predicted a net loss of 40 ns and instead measured a net loss of 59 ns.) Perhaps using the speed caused by gravity in combination with travel speed gives us a more correct method of determining total dilation.

It should also be noted that while Hafele and Keating proved both travel and gravitational dilation, they only achieved an accuracy within 10%. This suggests there may be room for some refinement of our understanding of these dilations.

With this new method of combining the two causes of dilation, we can proceed to evaluate orbits.

We can fill in our table with the following:

Radius R_s	Orbit Speed	Speed From Gravity	Total Speed	Total Dilation
0.50	1.000c	1.0000c	1.0000c	1.0000
0.75	0.816c	1.0000c	1.0000c	1.0000
1.00	0.706c	1.0000c	1.0000c	1.0000
1.25	0.632c	0.8944c	1.0000c	1.0000
1.50	0.577c	0.8165c	0.9448c	0.6724
1.75	0.534c	0.7559c	0.8746c	0.5151
2.00	0.500c	0.7071c	0.8184c	0.4253
2.25	0.471c	0.6667c	0.7714c	0.3636
2.50	0.447c	0.6346c	0.7335c	0.3203
2.75	0.426c	0.6030c	0.6977c	0.2836
3.00	0.408c	0.5774c	0.6681c	0.2559

Now we can see that matter cannot orbit at or near a black hole. It may fall into it and accelerate to light speed. Or if it attempts to enter an orbit, it will become fully dilated.

This data has been based on a black hole in which its mass is very concentrated at its center, as it would be for a singularity. But I don't believe black holes contain singularities or any object which is close to being a singularity. The reason is because the Heisenberg uncertainty principle doesn't allow it, as we saw in the early universe. I believe they are likely to contain an even density of matter. This arrangement will not affect the conditions surrounding an event horizon, but will change the effect of being inside it.

Even Density Black Hole

For discussion purposes, now, we'll imagine that within that perimeter, matter is distributed evenly. This means that the force of gravity becomes zero at its center and is strongest at the outer edge of it.

In order for this situation to persist the objects inside the perimeter must all rotate so that it doesn't collapse. These objects are

not all orbiting a black hole. They're orbiting something less powerful. Those at the center only orbit a small amount of mass. Further out, they orbit more mass and have a greater pull inward. Only at the perimeter could objects be said to orbit something of black hole force.

We can create a new table for orbits within this evenly distributed black hole. By determining the volume inside each radius, we can find mass and therefore gravitational pull at each radius. This leads us to orbit speed. We'll also calculate vibration speed from gravity and combine that with travel (orbit) speed. From this total speed we can determine total dilation.

A formula for travel speed may be developed from our original formula of

$$v = \sqrt{\frac{GM}{r}}.$$

We'll adjust our mass according to how much is contained within our radius. Since mass is evenly distributed in this black hole, we just need to measure volume differences. Our portion of the whole volume will be

$$\frac{\frac{4}{3}\pi p_s^3}{\frac{4}{3}\pi 1^3}$$

where

p_s = the portion radius we're using,
1 = the whole radius of our black hole.

This relationship simplifies to

Portion of the whole volume $= p_s{}^3$.

Now we can plug this in to adjust our mass and also put in our portion of the Schwarzschild radius in place of r:

105

$$v_t = \frac{\sqrt{Gp_S^3}}{\left(p_S \cdot \frac{2GM}{c^2}\right)}$$

This simplifies to

$$v_t = \frac{p_S c}{\sqrt{2}}.$$

Or, for speed as a portion of light speed

$$\frac{v_t}{c} = \sqrt{2}.$$

To determine our vibration speed we begin with $\sqrt{2GM/r}$. With the 2 in the numerator, we get the same answer as for travel speed, but without the $\sqrt{2}$:

$$\frac{v_v}{c} = p_S$$

Radius R_S	Orbit Speed	Vibration Speed	Total Speed	Total Dilation
0.0	0.0000c	0.0c	0.0000c	0.0000
0.1	0.0707c	0.1c	0.1157c	0.0067
0.2	0.1414c	0.2c	0.2315c	0.0272
0.3	0.2121c	0.3c	0.3472c	0.0622
0.4	0.2828c	0.4c	0.4629c	0.1136
0.5	0.3536c	0.5c	0.5987c	0.1845
0.6	0.4243c	0.6c	0.6944c	0.2805
0.7	0.4500c	0.7c	0.7864c	0.3823
0.8	0.5657c	0.8c	0.9259c	0.6223
0.9	0.6364c	0.9c	1.0000c	1.0000
1.0	0.7071c	1.0c	1.0000c	1.0000

Beyond *1.0 R_S*, the orbits are the same as for any other black hole.

From this data we can see that the interior of an even density black hole can have very mild gravity, low speeds, and low dilation. The only requirement to maintain this mild environment is that its contents be rotating. Without rotation, objects fall toward the center where interactions become intense. It's this pressure which seeks relief, and it eventually finds it by rotating.

Chapter 12 – Gravity is Soft

In discussing strong gravity there's often an assumption that an environment is either escapable or not. I believe the correct approach to these situations is not to consider whether the location is escapable but how far an object may get without a propulsive force.

The formula for escape speed is the same as the speed caused by gravity on an immobile object:

$$v = \sqrt{2GM/r}$$

Remember how this formula is derived from the formula for gravity combined with the formula for how far can be traveled with a certain amount of acceleration (or deceleration). It defines escape as being distance r when gravity is constant. But when leaving a gravitational source, the force of gravity decreases with distance at just the right rate to allow the object to slow but never quite stop.

Let's see this formula in action. Imagine I throw a ball upward at the escape speed of earth. The mass of the earth is 5.97×10^{24} kg and we're 6,378,000 meters from its center of gravity. This gives us an escape speed of about 11,200 m/s.

As the ball goes up, gravity slows the ball by 9.8 m/s each second (ignoring friction from air). At further distances gravity becomes less. The speed of the ball slows, but it continues to get further away where gravity decreases. The ball travels indefinitely, slowing and slowing but never stopping because there's always a little speed left and not quite enough gravity to take it all away. Had I thrown the ball slightly slower than escape speed it would eventually stop and fall back to earth. But equal to or faster than the escape speed it never returns, it just keeps slowing.

There's an old concept that light loses energy as it travels. This had been used to explain why light from distant galaxies appears redshifted, or of lesser energy. It's referred to as tired light. I don't believe this correctly explains why light from distant galaxies is redshifted, but the concept does have an application. It describes the softness of gravity.

Travel distance, or time, should not affect light. So light can't actually get tired from that. Traveling through space is essentially friction-free coasting from place to place. An object in empty space doesn't need any force added to it to travel any distance once it's moving. It just needs for nothing to slow it down. This applies to light also.

While photons may not encounter physical obstructions, there are accelerating forces along the way in the universe. Gravity not only bends light, it also attempts to speed it up or slow it down. Gravity can't change the speed of light, but it can pull on its waves because those contain the mass equivalency. This redshift can be thought of as tired light.

Just as a ball I throw into the air is constantly accelerated against by gravity, light waves are too. Gravity has a two-stage effect on light emitted from a source. The redshift of light trying to escape from gravity is the second one.

The first is that it's emitted at a lower frequency based on the dilation of the emitting source. This can be inversely calculated by the gravitational dilation formula. The formula tells us of the lost experience of the source. That lost experience results in longer waves emitted within longer periods of time. The source could be thought of as living a slow life. Essentially, an object which is dilated due to gravity, just like with speed, emits a lower level of energy while it experiences emitting a normal level.

For example, a person on earth with a ruby laser would measure it to emit 694.3 nm (nanometer) waves of light, while it actually emits a very slightly longer wave because of the gravity and travel of the earth. The same experiment done on the moon would result in less gravitational dilation, but more dilation from travel. Since the observers are in the same environment as the laser the change can't be measured. An observer receiving these beams of light in space, however, would see the dilation effect as he measures the wavelengths to be different.

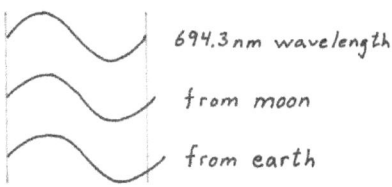

694.3 nm wavelength

from moon

from earth

The difference is very slight because these conditions are mild. But an observer with very little dilation because he's at rest in space a good distance from significant gravity should be able to measure the dilation, and therefore the movement or gravity of the earth and the moon.

Once emitted, the light has to fight against the gravity of where it's coming from as it travels. Just as a ball is continuously accelerated against after I throw it up, the light waves are too. This also means waves coming toward gravity get shortened by the same amount, just as the ball accelerates during its fall. Both the dilation of distant stars and constant gravitational pull cause color shifts and need to be considered when determining their location and movement. Stars at the far reaches of the universe could appear blue shifted to an observer at the center of the universe as the gravity of the universe pulls it inward, if there were no other effects.

The amount of acceleration against travel depends on the time it's exposed to the gravitational force. The faster I throw the ball, the more speed it has that it may give up, the more time it can be accelerated against, and the higher it can go before it reaches zero speed and turns around to come back, or eternally slows if it is thrown faster than escape speed.

This is the concept behind the inescapability of a black hole. The distance R_S is how far an object may get, traveling at light speed against the gravitational force at that radius, before its speed is reduced to zero. The real effect of gravity inside a black hole depends on the arrangement of matter and energy in it.

Light can travel indefinite distances, and gravity has indefinite reach. But light cannot slow down. Instead, its waves get longer. The more time a photon spends traveling away from gravity, the more gravity is able to slow, or stretch, its waves. This occurs according to the force of gravity where the photon is at. So it's a changing force as light moves away from gravity.

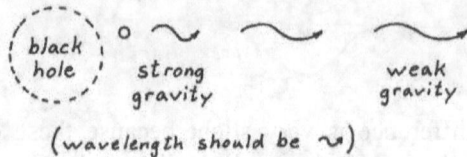

(wavelength should be ~)

Just as gravity will always pull a passing object into a spiral path no matter how far away it gets to be, the waves of departing light will continue to stretch for as long as gravity is pulling against them. No matter the distance, gravity will always continue to take a little speed away from (or lengthen the waves of) an object which is leaving it. If the initial speed is greater than escape speed it will continue to slow for eternity.

The reason there's always a limit for escaping gravity when traveling at less than escape speed is that gravity doesn't reduce speed by a portion. It reduces it by a quantity. If it was reduced by a portion, there would always be some left. A portion of a portion of a portion of... will always leave something.

Imagine that the sun was a black hole. We can determine how far an object could travel from inside if it departed at light speed. Since the sun's mass is about 2×10^{30} kg, its R_S would be approximately 3000 meters from its center.

We'll depart from just inside the perimeter and assume that all of its mass is concentrated at its center so that gravity continues to increase as we go further inside.

Starting Radius	Starting Speed (m/s)	Gravity (m/s²)	Time to Travel 2500 m	Ending Speed (m/s)
2500 m	300,000,000	2.135×10^{13}	0.000008 s	122,100,000
5000 m	122,100,000	5.338×10^{12}	0.000021 s	12,660,000
7500 m	12,660,000	2.372×10^{12}	0.000197 s	0
10,000 m		1.334×10^{12}		

112

This object lost all of its speed somewhere after 4500 meters away from the Schwarzschild radius. This is because it began inside the perimeter where the escape speed was greater than light speed.

This concept applies to all objects, including light itself, which are trying to leave any gravitational source.

It's this quantity reduction which makes all environments inescapable at slow enough speeds, and some environments inescapable even at light speed. If gravity provided a portion reduction, there would be no limit to how far any object could travel since it would never lose all of its speed. A ball thrown upward at any speed would never come down no matter how slowly it was thrown. It would eternally slow, appearing to be suspended in the air. All objects in elliptical orbits would never form an ellipse. At the far arc, the aphelion (the point on the orbit of a celestial body that is farthest from the sun), they would slow their outward movement indefinitely. They would never stop moving away so that they can fall back toward their attractor. No orbit would resemble an ellipse, just a continuous outward spiral. A cannon ball would never return to earth. It would seem to establish a low, slow orbit. Its forward motion would slow because of friction. Its upward motion would slow from friction and gravity, but never completely.

In the following table we see an object leaving the perimeter of our black hole at light speed and losing portions of its speed. Although it slows, it continues to travel indefinitely. I began by halving speed after 0.00001 second. Then I applied the inverse square law and found ever decreasing portion reductions.

Starting Radius (m)	Starting Speed (m/s)	Gravity (per 0.00001 s)	Time to-Travel 3000 m	Ending Speed (m/s)
3000	300,000,000	1/2	0.00001 s	150,000,000
6000	150,000,000	1/8	0.00002 s	112,500,000
9000	112,500,000	1/18	0.000023 s	98,200,000
12,000	98,200,000	1/32	0.000031 s	88,800,000
15,000	88,800,000	1/50	0.000034 s	82,800,000

As long as we reduce the speed by a fraction, there will always be some left no matter how slow our initial speed was or where it departed from.

This is the reason why light cannot escape a strong gravity source, such as a black hole. Enough force, given enough time, reduces any speed to zero, or eliminates the waves of light entirely. The photon, in theory, could always continue but its waves eventually get slowed to a halt. All of their speed is eventually taken away. Without the waves there isn't anything left. It must turn around.

When we consider whether light, or any object, can escape from a location we must qualify that with how far we consider escape to be. Because light speed is constant, this translates into a distance limit.

Light, or any object with momentum, can actually travel against any amount of gravitational force for some amount of time to reach some distance.

The answer to the question of whether light may escape from a black hole or any environment depends on where it departs from, and how far you want it to go. If it leaves the perimeter of a black hole, or just inside of it, it may be able to get out a distance. Matter behaves the same, so long as it doesn't become fully dilated and turn into light. Then it still behaves the same but isn't the object it started out as.

Singularity

The tired light concept limits how far anything can go from a center of gravity while traveling directly away. There is no event horizon. There is no one limit of escape from any gravitational body. Escape is a case-by-case issue.

But this doesn't mean the world will collapse. I don't believe the expansion of the universe will even slow. There's a reason why it's expanding.

Chapter 13 - All Orbits Accelerate

Absolute relativity predicts that all gravitational orbits accelerate, and as a result they also expand.

This is due to the effects of energy shift and the Doppler effect on gravity. And it's based on the absolute speeds of objects whether they're matter or energy. The result is that gravity causes objects in orbit to gain speed.

Energy Shift

When an object runs into energy, it receives that energy from a more forward direction than the energy was traveling. The direction, or angle of the energy seems to change. This occurs because both the energy and the object are moving. They both have speeds, so they encounter each other at an angle which is determined by those speeds.

Gravity is the interaction of two or more energy waves. The waves which give us the effect of gravity represent the total mass-energy of the object which they're a part of. And it travels at the speed of light, so it is pure energy.

These waves encounter each other the same way moving objects do. When a moving object runs into another moving object coming toward it, their angle of impact is more forward than their angles of travel.

For example, if two objects traveling at the same speed collide at 90° to each other, then they'll each respond to the impact at 45°.

If you were running at 10 mph past me and I threw a ball perpendicular to your path. You wouldn't catch it at your side. You would catch it at a 45° angle to your running, since you're running into it just as much as it's coming to you.

When matter runs into gravity, the wave which is the source of gravity is the faster moving of the two objects and the response to gravity is a pull toward the impact direction, not a push away. If an object could run into gravity at a 90° angle at light speed, it would be pulled equally forward and toward the direction the gravity came from.

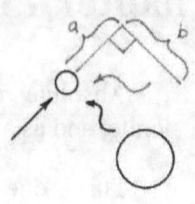

The force of gravity would be effective at an angle, but may be considered as divided into two vectors of pull.

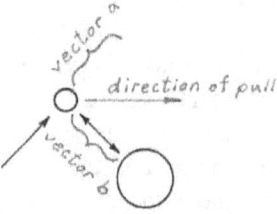

A planet orbiting a star at near light speed would experience only half of the star's attraction toward it. The other half of the gravitational pull would be in the direction of its orbit. The planet would be caused to orbit faster.

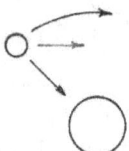

In an orbit, gravity is generally coming from an angle 90° to travel direction. Because of this we only need to consider the object's travel speed as it relates to light speed.

We can determine how much of the gravitational force will accelerate the orbit with $(v/c)/(1 + v/c)$.

And the amount of pulling toward the attractor (the star) is the remainder, or $1/(1 + v/c)$.

As an example, we can imagine a planet traveling at 30,000,000 m/s around its star (1/10 of light speed). Its speed (v) as a portion of light speed (c) is 0.1.

$$v/c = 0.1$$

So its acceleration in orbit will be *0.1/(1 + 0.1)*, or 0.0909, of total gravitational force.

Its pull toward the star is now only *1/(1 + 0.1)*, or 0.9090, of total gravitational force. It has given up some of its pull toward the star to gain speed in orbit. But that doesn't mean it'll drift away. The gravity the planet is running into gets enhanced.

Doppler Effect

Any object which is traveling toward an energy source will run into energy and experience more of it than if it was still. This is because, as absolute relativity tells us, all energy moves according to space itself not in relation to any object. Energy has absolute movement, not just relative movement.

If I threw you a ball every second at 10 mph and you were standing still, you would receive a ball every second at 10 mph. But if you ran toward me at 10 mph you would receive 2 balls every second at 20 mph each. Just as the frequency of a siren increases when it's coming toward you, how frequently you receive the balls increases.

At a right angle, you would receive the ball at the hypotenuse of your running speed and the ball's travel speed, and the hypotenuse of its rate too.

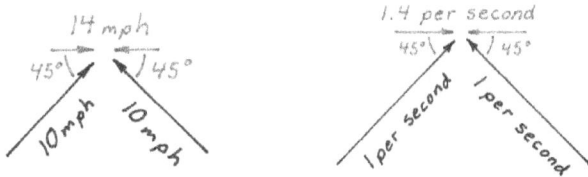

Receiving gravitational waves works the same way. If you were running into gravity, you would experience more gravitational pull.

We can add this factor to our example orbit. The increase in gravity is the same as the Doppler blueshift (increase) of *v/(1- v/c)*. We can apply it by multiplying our gravity by *1 + v/(1 – v/c)*.

117

At 0.1 of light speed travel, our total gravitational force increases by 0.111 to be 1.111. If we add this to our energy shift formula we get

$$gravity \left(1 + \frac{v/c}{1 - v/c}\right)\left(\frac{v/c}{1 + v/c}\right).$$

This gives us an increased acceleration factor of 0.101. And our attraction to the star is 1.010 of what it would be at rest.

The Doppler effect is able to accelerate the orbit while slightly increasing pull toward the star.

Migration

Readers who are familiar with absolute relativity might recall the concept of energy migration. Energy shift results in a migration of energy within any system, or collection of particles, which exchanges the energy.

Since gravity does not seem to be absorbed and re-emitted as other forms of energy (free energy), it should not migrate. If it remains a part of its emitting source, or returns to it, it cannot migrate. Migration only applies to transferred energy such as heat, light, or the photons exchanged between electrons which creates the flow of electricity.

Gravity appears to be an effect of energy which remains part of its source. Gravity waves are known to be either continuously or regularly emitted by their particle. But they behave as if they are bound to it and can't migrate.

The shift of direction from running into gravity, combined with its enhancement by the Doppler effect, increases orbit speed while providing a slight increase in attraction. Regardless of orbit speed there will always be an accelerating effect. This will cause an outward spiral. A balance between the two found by way of the outward spiral cannot be avoided.

If half of the effect is applied to increasing the speed of orbit and the other half is applied to expanding the size of it, then the period remains the same. The time it takes for each orbit is unchanged. This creates difficulty in detecting the effect.

When watching orbits from great distances, sizes can be hard to determine. But frequency of the orbit is simply the time between a redshift as an object moves toward us and the blueshift of it moving away during its orbit. If this orbital period remains constant, we might assume its speed and orbit size occur according to the standard formulas for stable orbits and calculate them that way.

Also, for objects orbiting closely there may be other effects hindering their acceleration, such as their atmospheres, tidal forces, or spin.

This acceleration of orbits may be hard to measure.

Consequence

The result of accelerating orbits is that they are forced to spiral outward. Gravity will always accelerate any orbit of less than light speed unless acted upon by another force.

One might think to correct for this by directing the orbiting object inward. But any push inward from circular would cause the satellite to be flung out of orbit and away from its attractor in less than one cycle. This would occur in the same way as orbit acceleration.

We saw how the Doppler effect causes an increase in how frequently energy is run into when traveling toward it. Likewise, it causes a decrease in how frequently energy is encountered when traveling away from an energy source. This means that objects traveling toward each other are more attracted to each other than they would be if they were going away.

Consider the planet orbiting the star we examined earlier. The planet started with a circular orbit and was accelerated into an outward spiral.

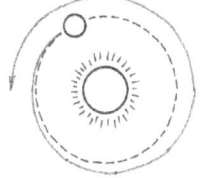

Let's begin again with a circular orbit and push the planet inward at a slight angle. As it runs into gravity waves at a more direct angle than previously, its speed increases faster than before. With this greater speed it has more centrifugal force than it acquired in the circular orbit.

Now its orbit is forced to expand even more dramatically than before. As its angle shifts from toward the star to away from it, the Doppler effect reverses. It feels less gravity from the star than it normally would at that distance and the speed it gained initially is not entirely taken away from it.

Within one cycle, the planet pushed inward is now further from the star than it started out. It's moving faster and is spiraling outward more dramatically.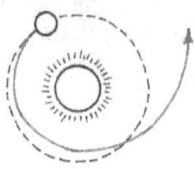

(I'm ignoring the eccentricity of orbits for simplicity. Because we know orbits aren't circular but elliptical, it's the ellipse which grows larger with each cycle. This still creates a spiral.)

It may not seem possible that all orbits could actually be outward spirals because we don't see it in our solar system. So let's look at a real example of an orbit we believe to be stable.

Earth

In our previous example, I gave the planet a very fast orbit speed. Earth doesn't travel nearly that fast. It only goes 66,900 mph around the sun. (While evidence shows us the earth is not moving this fast, the movements of our solar system are most easily understood by describing it as orbiting the sun.)

If we apply orbit acceleration to earth we find that it should increase in speed and orbit size by 00.03% each year. This is not an amount to be concerned about for our survival, but I also don't believe this is being noticed. The reason may be related to us orbiting within the atmosphere of the sun, influences from other planets, tidal forces, and spin.

120

Based on this behavior we know that any object which approaches but does not collide with another more massive object will gain speed from the interaction. It will depart the gravitational influence with more speed than it entered with. This is a source of acceleration in many gravitational encounters.

Having looked at our own orbit, we can see that it is not unreasonable to suggest that all orbits accelerate and result in outward spirals. The concept that there is no unchanging orbit doesn't have a concerning impact on our foreseeable future due to our slow travel speed. But it's useful in examining faster moving systems and predicting the universe as a whole.

We can see that gravity causes rotating systems to accelerate and that causes them to expand. Because of this, the attractive force of gravity creates the effect of a repulsive force.

I believe this can answer some questions about our universe.

Chapter 14 - Consequences of Acceleration

The effects of absolute relativity applied to gravity may be the cause of orbits accelerating. If it is, then this can solve many mysteries that exist today. I'll begin with what I believe is the simplest and proceed toward the conclusion of our universe.

Dark Matter

It is commonly believed that each galaxy may be surrounded by a halo of dark matter, a spherical collection of matter of almost 9 times the visible mass of the galaxy. And this matter is only evident by its gravitational effect. It doesn't interact with light or other radiation or matter.

Energy, too, does not interact in obvious ways with light. But it must travel and could not remain in orbit around a galaxy unless that galaxy was smaller than Gm/c^2.

I believe that dark matter is not matter or energy, but a phenomenon. It's the orbital acceleration caused by gravity.

In the early part of the 20th century, Jan Hendrik Oort noticed that stars in our galaxy orbit faster than they should for how far out they are. In 1933, Fritz Zwicky noticed a similar behavior. He found that galaxies in clusters orbit their cluster faster than can be accounted for by visible mass. He proposed that there may be another form of matter which we can't see. And in 1970, Vera Rubin speculated of the existence of unseen mass when she calculated that the outer stars of the Andromeda galaxy are orbiting just as fast as the inner ones, too fast for the mass of the galaxy. It seems that excessive orbit speed is common.

My explanation for this is that the acceleration of orbit speed is caused by gravity. As I pointed out, the pull forward by orbiting bodies is caused by energy shift. And the enhancement provided by the Doppler effect contributes to this acceleration. This increases centrifugal force which causes the orbit to expand. The result is both a continually increasing speed and an expanding orbit. And, as with any acceleration, the longer this goes on the more dramatic it becomes.

Converting Matter into Energy

As objects orbiting within any gravitational system are accelerated and spiral outward, their speed will eventually reach that of light. This is because gravity has indefinite reach. It is always present, so it can always provide more speed.

I suggested previously that gravity doesn't accelerate objects by giving them energy, but converts matter into energy. This is how gravity's acceleration is able to remain constant and not be inhibited by mass or increasing mass equivalency of the object it's accelerating. As gravity converts matter into energy, its only limiting factors are the speed of light and the full conversion of all of the mass of the object. Once an object becomes 100% energy it must travel at light speed since it is light. Beyond that, gravity can only increase the frequency of that energy.

I believe that any matter which happens to be orbiting inside an extreme gravitational environment, such as a black hole, is being accelerated to light speed, and turning into energy just as the matter orbiting outside of it is.

Identity Crisis

Gravity converting matter into energy causes particles to lose their identity. A particle can't always bring its properties into a black hole. And it certainly can't convert into pure energy with them. Particle properties, such as electric charge, are conserved. They can cancel by opposite charges coming together, but one alone can't just disappear.

As a particle is accelerated to light speed most of its mass may convert into energy, but some must remain as matter to carry properties. For example, a proton has 1837 times the mass of an electron or anti-electron. It also has an electric charge which can only be carried by matter. While most of a proton's mass may convert into energy, it must leave behind an anti-electron to hold its positive charge. The anti-electron will be attracted to and eventually find an electron (negative charge) to annihilate with. The charges will cancel and only energy will remain.

Most of our matter exists as protons and neutrons. When they decay, they release two X-ray photons along with any small particle necessary to carry charge, as well as neutrinos which have very little mass.

Dark Energy

Our universe is said to expand due to a force which is from an unknown source. We call this unknown source dark energy. I believe this is not energy at all, but is an effect of gravity behaving according to absolute relativity.

Dark energy, also known as the cosmological constant, was speculated to exist in order to explain the expansion of the universe. Since all the matter and energy of the universe is attracted to each other by gravity, it would be easy to guess that it should eventually collapse unless another force acts against it.

In a big bang explosion, inertia can serve this purpose to some extent. However, inertia is a tendency to a constant speed, and gravity is an accelerating force. If inertia was the explanation for expansion, our universe would have been slowing its expansion due to gravity since it began, until it ultimately reversed and collapsed.

The exception to this is said to be that both speed and gravity diminish equally during the expansion, that they are perfectly balanced. In that case, a forever slowing expansion is said to be possible. This is the explanation given by some scientists who calculate the density of the universe to find this balance at what they refer to as $\Omega = 1$. This would mean the universe expands exactly at its escape speed.

A long-lived universe is said to require a very precise balance, even to keep us expanding as long as we have. This is the equivalent of throwing a ball up into the air just barely fast enough that it never falls back down.

An obstacle to this concept arose in 1998 when Adam Reiss announced that expansion is speeding up, not slowing. It seems there is a force, or the effect of a force. This appears to definitively eliminate the possibility of inertia alone keeping expansion going.

My suggestion is that the universe is rotating, and with all galaxies orbiting the center of the universe, they too accelerate according to predictions of absolute relativity just as stars do in galaxies, and galaxies do in clusters.

Some may believe this rotation does not occur, as there is no evidence of it when comparing reference frames such as the cosmic microwave background radiation or galaxies' relationships to each other. In response to this, I suggest that all matter participates in this rotation. So far, it seems measurements have only been taken relative to other matter. I believe that by measuring relative to absolute space we can determine true movement.

A further difficulty in determining rotation is the possible rate. If the universe is 13.8 billion light years in radius the fastest it could rotate is once every 81 billion years because it is limited by light speed at its outermost limit. The furthest galaxies would be approaching light speed. Inner parts would move very slowly.

Even a much smaller universe size would have a rotation that is difficult to measure.

Chapter 15 - Limit of Universe Expansion

We began our exploration of the universe by considering how it could have come into existence. And it appears to have occurred by probable distribution from a size too small to be sustainable. But even at a comfortable size, as soon as the particles began interacting, many smaller unsustainable situations must have occurred. Areas became too dense and needed to re-distribute their particles. This is because of the initial particles having energy expressed as speed. The result was a chain of explosions beginning with one large one and followed by explosions of decreasing size. This would force the initial cloud of the universe to grow larger. But we don't know how large.

Movement theory tells us that this quantum tunneling is actually common movement. It's how everything travels. And travel obeys the speed limit. This means the rate of growth of the universe could not have been more than the speed of light.

While it's difficult to determine the size of the universe, we have been able to determine its age using the decay of elements. And we estimate it to be about 13.8 billion years. This fixes the limit for the size of the universe at about 13.8 billion light years, since it is the furthest matter could have reached in that time.

Having estimated the total mass of ordinary matter and energy of the universe, we can confidently say that this mass must be within the Schwarzschild radius. The universe qualifies as a black hole. And we've found that any attempt to escape a black hole is likely to turn matter into energy.

As gravity is soft and travel from inside a black hole is limited, any energy which tries to leave will be brought back eventually. And, in time, it should find itself in orbit in a photon sphere of 1/2 of the Schwarzschild radius.

With matter at the outer reaches of the universe being converted into energy to accumulate at 1/2 of R_S, I believe this is a likely size for the bulk of a collection of matter such as the universe.

(Some matter may still exist at any distance beyond this radius but less than the distance it could have traveled at light speed.) While this is a small size, I believe it's possible because of how deceiving distances can be when applying absolute relativity.

Additionally, it may not be as small as we expect. Seeing how energy should accumulate in this orbit leads to an additional question. How much energy may be in orbit there already, and might that contribute to the gravity of the universe? The size of that orbit is based on the matter and energy we find in the universe plus the energy in orbit. Any energy already in orbit is undetectable as it does not affect us as long as we're inside it. Leaving the energy sphere, gravity would become more dramatic based on how much energy is already in the sphere. For now, we'll leave this unanswered.

With the orbit of energy as our size, let's look at the environment inside at various distances from its center. This is similar to the even density black hole we looked at earlier, but may have more intense orbits since it's smaller. We can calculate it in much the same way, but instead of the Schwarzschild formula for radius we'll use the formula for the energy orbit. Our orbit speeds are the same as the portion of the energy orbit as a portion of light speed.

$$\frac{v_t}{c} = p_E$$

Vibration speed as a portion of light speed is

$$\frac{v_v}{c} = p_E \sqrt{2}.$$

Radius (R_E)	Orbit Speed	Vibration Speed	Combined Speed	Dilation
0.1	0.1c	0.1414c	0.1637c	0.0135
0.2	0.2c	0.2828c	0.3273c	0.0551
0.3	0.3c	0.4243c	0.4911c	0.1289
0.4	0.4c	0.5657c	0.6547c	0.2441
0.5	0.5c	0.7071c	0.8184c	0.5746
0.6	0.6c	0.8485c	0.9821c	0.8116
0.7	0.7c	0.9900c	1.0000c	1.0000
0.8	0.8c	1.0000c	1.0000c	1.0000
0.9	0.9c	1.0000c	1.0000c	1.0000
1.0	1.0c	1.0000c	1.0000c	1.0000

Our Place in the Center

The Hafele and Keating experiment shows that we experience no significant dilation which may be attributed to the gravity of the universe.

Based on popular estimates of the size and mass of the universe, we should be fully dilated by its gravity regardless of our location in it. This cannot be overcome without great adjustments to the mass, size, or distribution of matter, or elimination of matter. It also cannot be overcome by the manner in which the universe expands, as we know gravity has unlimited reach.

Our lack of gravitational dilation indicates that we're located at the gravitational center. Meanwhile, our lack of travel speed tells us we're at the expansive center of the universe.

Applying Gm/c^2 to our universe, and putting all of its mass into it, we can determine that its orbit for energy should be at a radius of 6.42 x 10^{24} meters, or 679 million light years. This is where the photon sphere forms, assuming the universe contains 8.65 x 10^{51} kg of total matter and energy. (I've eliminated dark matter and dark energy from these figures. If they were included, the total mass would be 1.73

x 10^{53} kg and the photon sphere would be at a radius of 13.5 billion light years.) This is dramatically smaller than the many billions of light years away which visible galaxies appear to be. But if there is energy already in the photon sphere, then its radius could be much larger. This is because the light in orbit is not only attracted to the mass within its orbit but also to the other light in the orbit.

In my book on absolute relativity I reconstructed a hypothetical universe using the rate of expansion as currently calculated. I found that if our universe has been expanding for 13.8 billion years and appears to be 13.8 billion light years at its furthest extent, it could only be about 6 billion light years in actuality. And the dimness and redshift of galaxies is being exaggerated by the effects of absolute relativity.

Since redshift is currently the main quality used to determine where a distant galaxy is and how it's moving we'll focus on that. My hypothetical data was as follows.

Galaxy	Apparent Distance (billions of Light Years)	Observed Redshift
1	1	0.0747
2	2	0.1495
3	3	0.2242
4	4	0.2989
5	5	0.3737
6	6	0.4484
7	7	0.5231
8	8	0.5979
9	9	0.6726
10	10	0.7473
11	11	0.8221
12	12	0.8969
13	13	0.9716
14	13.8	1

(Redshift was calculated using Hubble's constant of 73.04 km/s per 3.26 million light years.)

For this demonstration I had only considered that each galaxy was moving directly away from us. This means my figures were minimum speeds and maximum distances.

Now, considering the necessity of all objects to spiral outward due to the effect of gravity, I believe we must admit that every galaxy is not moving directly away from us but is spiraling outward at some angle.

The effect of these spiral paths is that the galaxies are not moving away from us as fast as they appear to be. Their redshift may be caused primarily by energy dilation from their total speed, which includes their spiral path and gravitational vibration.

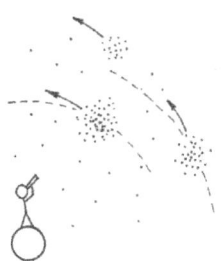

If the universe is rotating and we attribute the redshifts we see to dilation from both orbit speed and gravity, we can determine the speed and distance of each galaxy.

From the observed redshift we can determine the speed which causes it by dilation. From that speed we can find the orbit it should be at within an even density photon sphere radius.

Redshift is a result of dilation, so *1 − redshift* is experience. We can enter that into the dilation formula as

$$v/c = \sqrt{1 - (1 - redshift)^2} \ .$$

This tells us speed as a portion of light speed. It will be the total speed of the galaxy (travel and vibration combined).

Galaxy	Observed Redshift	Combined Speed
1	0.0747	0.3792c
2	0.1495	0.5260c
3	0.2242	0.6310c
4	0.2989	0.7131c
5	0.3737	0.7796c
6	0.4484	0.8341c
7	0.5231	0.8790c
8	0.5979	0.9156c
9	0.6726	0.9449c
10	0.7473	0.9675c
11	0.8221	0.9840c
12	0.8969	0.9947c
13	0.9716	0.9996c
14	1.0000	1.0000c

With total (combined) speed, we can find our orbit radius by reversing the combined speed formula.

The original combined speed formula is

$$combined\ speed = 0.945 \sqrt{v_t^2 + v_v^2}\ .$$

To determine orbit radii within an even density photon sphere radius we use

$$\frac{v_t}{c} = p_E$$

and

$$\frac{v_v}{c} = p_E\sqrt{2}$$

where
v_t = travel speed in m/s,

v_v = vibration speed in m/s,
p_E = portion of the photon sphere radius,
c = the speed of light as 299,792,458 m/s.

We can assemble these formulas to determine combined speed based on portion of radius of orbit (p_E) :

$$combined\ speed = 0.945\sqrt{p_E{}^2 + (p_E\sqrt{2})^2}$$

To solve for portion of the energy orbit radius (p_E) we have

$$p_E = \frac{v/c}{0.945\sqrt{3}} \ .$$

This tells us that the orbits of our visible (hypothetical) galaxies are:

Galaxy	Radius (R_E)	Radius (Light Years x 10^6)
1	0.2317	157
2	0.3214	218
3	0.3855	262
4	0.4357	296
5	0.4763	323
6	0.5096	346
7	0.5370	365
8	0.5594	380
9	0.5773	392
10	0.5911	401
11	0.6012	408
12	0.6077	413
13	0.6107	415
14	0.6110	415

This is based on our universe having a photon sphere at 679 million light years. Gravity and orbit speed cause great dilation at the outer reaches converting matter into energy. Matter, then, only exists within 415 million light years radius.

I suggest that the contents of the universe are spiraling outward, accelerating to their end. At that point, as energy which is unable to escape, they join the photon sphere in orbit.

Lifespan of the Universe

The speed of light seems to be the key factor in how fast everything happens. It sets the pace for all activities of the world. It is THE speed for movement at the smallest scale, according to movement theory.

It's also involved in determining the size of gravitational systems, which then leads to how long it will take them to complete their life. I believe all gravitational systems eventually convert their components into energy. The conversion is the end of the cycle of matter, and it's the destiny of our universe. The time it takes for gravity to take everything from the initial cloud of particles to full conversion into energy is determined by how far away the end is. The end of material existence occurs at the edge of the universe.

The size of the universe is not just determined by light speed, or how fast things happen. The other factor involved is how much stuff there is in it.

If the size of the universe is determined by Gm/c^2, the radius of an energy sphere (or even the Schwarzschild or other similar formula), then mass is the final determining factor. The distance across the universe is directly proportional to mass. The amount of mass in the universe brought our world into existence. And it also determines how long it will remain.

If our sun and 9 planets were all that existed, then the universe would have been much smaller than our solar system. It would have only grown to about 3 kilometers in diameter. And with activities occurring at speeds dictated by light speed, it would complete its life very quickly. The lifespan of the universe would have been the blink of an eye. Matter would come into existence and convert to energy before complexity could form. No elements could have fused.

Billions of years are needed just to form the elements necessary for life. Then an environment suitable for it needs to be established.

The number of stars in the sky determine how long the universe will live. So each star could be said to contribute an amount of time for our world to accomplish any purpose it has. Billions of years of development occurred before the window for biological life opened. And after it closes it will take the universe billions of years to wrap up its existence.

Chapter 16 - The Beginning and the End

I believe that when we look deep into space we see both the beginning and the end of creation, that they are occurring simultaneously at the edge of our material world. It's true that we're looking back in time, but I don't believe it's so far that we can't say that the beginning and end aren't both occurring together and continue even today. This can be explained by the universe being a gravitational environment in which gravity both slows the assembly of matter and causes its conversion into energy.

We receive background radiation which is evidence of both, and this radiation is consistent in all directions. It is primarily in the form of microwaves, X-rays, and neutrinos. These three types of particles tell us of the assembling of atoms and the breaking apart of the pieces of atoms. I suggest this activity all occurs at the edge of our material world.

Microwave Background

No matter which direction we turn there are low energy waves coming at us. And their distribution is very consistent.

George Gamow, Ralph Alpher, and George Hermann predicted this cosmic microwave background radiation as a remnant of the very early universe. It is expected to be from the initial hot cloud of matter emitting microwaves of 0.0048 mm wavelength. This radiation is released by helium atoms as they form. When the initial cloud of material had cooled enough, electrons were able to join helium nuclei. And this joining process produces the microwave photons. This is predicted to have occurred 379,000 years after the big bang.

In 1965, Arno Penzias and Robert Wilson detected this radiation with a radio telescope at Bell Laboratories in New Jersey. Since then, we've studied its slight variations which seem to tell us of how particles originally began to gather in order to form galaxies.

We receive this energy not at its emitted 0.0048 mm wavelength but at 5.28 mm. This is believed to be due to the waves being stretched by the Doppler effect as this cloud moves away from us very fast.

If the cloud moved directly away from us, as is expected, then this stretching of wavelength tells us the cloud was traveling 99.91% of light speed (redshift is change in frequency divided by frequency, $\Delta f/f = v/c$) when the microwaves were emitted. And if the waves were emitted that far back in time, then they traveled 13.8 billion light years to reach us. So they must have been emitted 13.8 billion light years away from us in every direction. The universe had to have been that large when it began and today could be 27.6 billion light years in radius. That doesn't describe the small hot beginning we believe we had.

Space Growth

Many scientists suggest the way this is possible is that the cloud of matter did originate very small and space itself grew. The dust, galaxies, and all other material has not actually been moving away from us. They just acquired more distance. And somehow, while they're actually at rest in space, they're not falling toward each other.

We can visualize this situation as two balls resting on a rug. The rug is space. If space is growing, then we can imagine the rug growing between them. This way they're each at rest on their part of the rug. This would be similar to them being on their own rug and their rugs being pulled away from each other. The balls, then, become further apart. They're at rest on their part of space, but become further from each other.

However, we know objects aren't fixed in space. They're free to move around, as we experience in our world. The balls aren't attached to the rug. And we have gravity. Everything's attracted to everything else. Given the opportunity, everything will move closer. We can represent this by putting each ball on a slope. In this way the balls tend to roll down the slope toward each other.

We pull on the rugs to expand space and create more space between the balls. But the balls fall toward each other. So we need to pull on the rugs fast enough to get them to be further apart. We have to pull them at the rate of expansion plus their rate of fall.

The problem we soon encounter is that gravity is an accelerating force. The balls roll toward each other faster all the time. So we need to pull the rugs faster all the time.

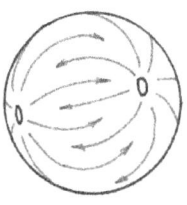

To solve this problem of needing to grow at an ever-accelerating rate, we could wrap space around into a ball as Bernhard Riemann suggested in 1854. While it's difficult to imagine our three spatial dimensions wrapped around like that because it doesn't seem to represent our world, this is a popular explanation. In this way, every direction you could go would take you back to where you started, if you went far enough. With this arrangement all objects not only pull on each other, they pull each other in all directions at once.

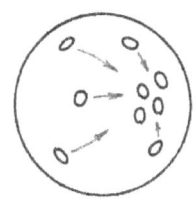

This solves the issue of needing accelerating space growth, but still leaves the galaxies free to pull together into a massive cluster somewhere in the universe. Galaxies are not evenly spaced, so they are bound to find preference and move that way.

In addition to these obstacles, there are other problems.

Suppose this dust cloud at rest did grow to be 13.8 billion light years away in all directions because space itself grew. In this way the dust emitted this radiation closer to where we are now, and the creation of space stretched its waves. So the microwaves, although they are light, didn't travel at light speed to reach us. They traveled at 600,000 mph (0.09% of light speed). This is much slower than the same radiation our microwave ovens use to cook our food, which travels at 670,000,000 mph.

In this model we have a varying speed of light and two different ways of traveling, movement as we're familiar with and space growth. This variation of speed may not be detectable, though, since perhaps time could grow according to distance also. So, galaxies are somehow held against falling toward each other from their position of rest, and space and time grow due to an unknown cause.

An alternative is to keep space fixed, and consider that cosmic bodies move through it. The world simply is as it appears to be. If we accept that all objects which change location, or distance from us, are actually moving, then the furthest back in time we can see is half of history. We can estimate this to be 6.9 billion years if the universe began 13.8 billion years ago. If all parts of the original universe were once together and some traveled away from us at near light speed, then today they would be 13.8 billion light years away. But the radiation we receive from them today would have been emitted 6.9 billion years ago when they were only halfway there.

This means the primordial helium formation occurred 6.9 billion years after the big bang, not 379,000 years as is expected. It would seem that the microwave background radiation is not from creation. And that it's not from the initial formation of the universe. But I believe it is. I believe this can be explained by time dilation.

The faster an object travels the less time it experiences. Fast moving objects exist in slow motion. If we were to travel for 6.9 billion light years at 99.91% of light speed, we would only experience 293 billion years. Additionally, gravity would cause a loss of time. And since this dust is expected to be at the outer reaches of the material world, it would experience the full gravity of the universe, and the dilation it causes. This dilation is very helpful for explaining the

140

microwave background, but I believe there are more adjustments to be made.

One is that if gravitational acceleration caused expansion of the universe, then the dust which formed helium didn't travel straight out from its origin. It spiraled. So it never got to be 6.9 billion light years away. Depending on the angle of the spiral, it could be emitted from much closer and therefore much more recently. If our universe's photon sphere is only 679 million light years away, the microwaves likely originated less than that distance away. So they were emitted less than 679 million years ago.

Once emitted, the microwaves must travel at light speed in a direct path to us for us to receive them. They can't spiral to us along the same path the cloud spiraled away in, so the Doppler effect is greatly reduced.

Now we have a cloud which has traveled in a spiral for 13.8 billion years to reach a distance of less than 679 million light years. Assuming its spiral speed is 60% of light speed (the speed of orbit just before turning into energy), its speed going away from us may only be an 8.5% increase in wavelength. If we know that the wavelength we receive is 1100 times the original length, then we still have almost the whole effect to account for. This can be accomplished by the dilation of the dust itself.

If the helium was only experiencing 0.00083 of absolute time and distance at the moment it formed, it would emit 5.8 mm waves instead of the 0.0048 mm which it normally would. The helium was 99.917% dilated at the time the waves were emitted. This should be the result of dilation from gravity and travel speed together.

I suggest that the microwave background radiation we receive today was emitted less than 679 million years ago from less than 679 million light years away by helium which was spiraling outward from us, at about 60% of light speed.

The creation process which began with the big bang continues to take place by dust which is so dilated that it doesn't know 13.8 billion years have passed.

X-ray and Neutrino Background

In addition to the microwaves, we also receive X-rays and neutrinos equally from all directions.

X-rays and neutrinos are created when matter is destroyed, or converted into energy. This commonly occurs at the event horizon of a black hole. Matter approaching a black hole gets drawn into a fast moving orbit at the event horizon forming what's known as an accretion disk. It is essentially waiting its turn to accelerate into energy.

The force of gravity at the event horizon converts matter into energy by accelerating it to light speed as it nears the black hole. This is done by travel speed in combination with vibration speed gained by being pulled by gravity. Their combined speed converts matter fully into energy. In order to achieve this full conversion, matter must give up any properties it has which cannot be carried by pure energy. Most significantly, it must give up electric charge. Only matter can have that.

Much of the quantity of matter becomes X-rays when particles break apart. A small amount of mass becomes a large number of neutrinos. They are the lightest particles of matter. And they travel near, but not quite at, light speed in order to maintain their small amount of mass (less than 3.9×10^{-30} kg). They are remnants of the shredding of larger particles and they're everywhere, like very fine dust. It's estimated that trillions of neutrinos pass through our bodies every second.

Our X-ray and neutrino background is expected to come from the destruction of matter around black holes. And their consistency is explained as possibly a blurring of many of these sources.

I agree that the conversion of matter is the process generating this radiation, but I don't believe it's coming from all of the black holes. If it was, I think we should detect concentrations based on the locations of galaxies. With wavelengths in the range of billionths of a meter, I believe the X-rays we receive should give us an image with enough resolution to determine this. And if space was expanding, we should also be able to measure a redshift of the X-rays since the galaxies are expected to be moving away from us at different speeds based on their distance from us. I don't believe this has been reported.

I believe the bulk of the X-rays and neutrinos may be coming from just one source which would have no significant Doppler shift. It is the acceleration of matter at the outer limits of the material universe. The inside of our universe's energy horizon.

As matter spirals outward from the center of any gravitational system it will accelerate, eventually reaching the speed of light. This should cause the same destruction of matter that we see outside of a black hole. It could happen at the outer limits of any gravitational system which has developed enough speed. It could happen at the edge of a massive or fast-spinning galaxy.

I believe this background radiation we find all around us is mostly caused by the rotation of the universe. At the outer limits, matter should be accelerated toward light speed. And what would be out there is the earliest parts of the initial universe which never got to fully form as the central parts of it did. It is the initial cloud of particles from the big bang. Now it's being destroyed.

The microwave background is evidence of the beginning of our universe, the creation of helium continuing to occur. And the X-rays are evidence of its end, that same matter finally transitioning into energy to eventually reach its destination at the energy horizon of our universe: the photon sphere.

It seems we're witnessing the beginning and end of our universe occur simultaneously. Gravity is accelerating all matter, causing matter to move further from each other and convert into energy.

This acceleration increases the rotation of the universe in which the outer parts are moving as fast as possible, while the center moves imperceptibly. This speed causes dilation so that the outer edge of the material world experiences almost no time. It exists in slow motion due to its high speed. In the center we get a full experience of time and witness the true age and history of the world.

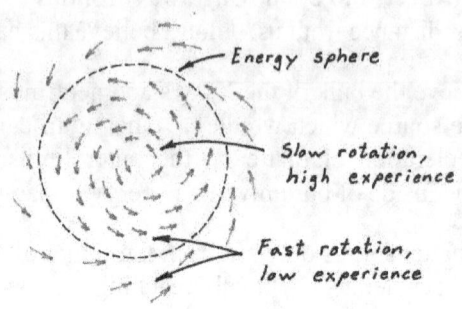

Energy sphere

Slow rotation, high experience

Fast rotation, low experience

The ultimate destiny of all matter is to be converted into energy.

Earth, too, will end this way. Even if it's at the very center, the effect which causes the rest of the matter to accelerate should cause earth to also. In the distant future, whatever matter is left alone at the center of gravity will cool to the point where uncertainty could relocate it. But that doesn't save it from destruction. That places it outside of the center so that it may be drawn back in and accelerated into energy in the process. If no other effect causes our matter to decay into energy, we should be accelerated to light speed by gravity and join the halo of light.

Chapter 17 - Conclusion

As we pursue the most fundamental workings of our world, I believe we should find fewer laws, not more. I believe there will be a consolidation of physics. By exploring gravity along with the effects of absolute relativity I think we can begin to see some simplification. If what I've presented is correct, then the Heisenberg uncertainty principle may be a fundamental law. And the law of conservation, also being a fundamental law, may be the last one put into place. This is because the declaration of a fixed amount of energy and matter seems to result in the big bang.

I refer to the universe as ours, not to suggest that there are others, but that it is for us. We appear to have the most special location in all of it. And its size, and therefore lifespan, seems tailored to suit us.

The universe is a very big and energetic place, and we are very small in it, both with our physical size and our lifespan, not just as individuals but as humanity. But we seem to fit right into the calm center of it as in a nest. There are many details to be figured out and likely revisions to be made. But I believe this structure can guide us to advancing our knowledge of it.

About the Author

Timothy Michaels is an artist and writer. Being a technically minded, "cerebral" thinker, he enjoys creating realistic art as well as exploring the world around him through physics. This led him to an explanation of color relationships using physics, which resulted in the development of his Color Calculator. Mr. Michaels has an exceptional gift for logic and an insight into the workings of the universe.

His scientific ideas rely on established theories and evidence, which are explained in a simple manner for general readership. Mr. Michaels provides new ways of understanding these theories. His ideas go beyond current science to solve problems and make new predictions. In doing so they provide the concepts and formulas necessary to advance science. Students of physics, astronomy, cosmology and other fields will find his explanations enlightening and his new ideas worth investigating.

His artwork may be viewed at:

www.tmsartgallery.com

Readers are invited to comment by sending an email to: 101timsplace@gmail.com. Be sure to put the book title in the Subject of your email.

Other Books by Timothy Michaels

Absolute Relativity: How Newton and Einstein Agree

Out of This World: The Movement Dimension

The Physics of Color Harmony

How We See Art

Walking the Cards: A Unique Drawing Method

Bibliography

Brooks, Michael. *13 Things That Don't Make Sense: The Most Baffling Scientific Mysteries of Our Time*, Vintage Books, 2008.

Chown, Marcus. *Solar System: A Visual Exploration of the Planets, Moons, and Other Heavenly Bodies That Orbit Our Sun*, Black Dog & Leventhal Publishers, Inc., 2011.

Clegg, Brian. *The Universe Inside You*, Icon Books Ltd., 2012.

Einstein, A., Lorentz, H.A., Minkowski, H., Weyl, H. *The Principle of Relativity*, Dover Publications, Inc., 1952.

Fleming, Thomas A. (editor). *Stars*, Cognella, 2011.

Graney, Christopher M. *Setting Aside All Authority: Giovanni Battista Riccioli and the Science Against Copernicus in the Age of Galileo*, University of Notre Dame Press, 2015.

Gribbin, John. *Einstein's Masterwork: 1915 and the General Theory of Relativity*, Pegasus Books Ltd., 2016.

Gunther, Leon. *The Physics of Music and Color*, Springer, 2012.

Henderson, Linda Dalrymple. *The Fourth Dimension and Non-Euclidean Geometry in Modern Art*, Massachusetts Institute of Technology, 2013.

Impey, Chris. *Einstein's Monsters: The Life and Times of Black Holes*, W.W. Norton & Company, Inc., 2018.

Kaku, Michio. *Physics of the Impossible*, Anchor Books, 2008.

Menzel, Donald H. (editor). *Fundamental Formulas of Physics*, Dover Publications, Inc., 1960.

Mercier Claude. *Calculation of the Mass of the Universe, the Radius of the Universe, the Age of the Universe and the Quantum of Speed*, Journal of Modern Physics, Scientific Research

Publishing, Inc., July 19, 2019.

Michaels, Timothy. *Absolute Relativity: How Newton and Einstein Agree*, Barnes & Noble, 2023.

Nassau, Kurt. *The Physics and Chemistry of Color*, John Wiley & Sons, Inc., 1983.

Oakley, C.O. *The Calculus*, Barnes & Noble, Inc., 1957.

Odenwald, Sten. *Imagining Our Infant Universe*, Astronomy, Vol. 50 Issue 4 April 2022.

Panek, Richard. *A Cosmic Crisis*, Scientific American, March 2020.

Randall, Lisa. *Warped Passages*, HarperCollins Publishers, 2005.

Rovelli, Carlo. *General Relativity: The Essentials*, Cambridge University Press, 2021.

Sheldrake, Rupert. *Seven Experiments That Could Change the World: A Do-It-Yourself Guide to Revolutionary Science*, Riverhead Books, 1995.

Susskind, Leonard. *The Cosmic Landscape*, Little, Brown and Company, 2006.

Van Zyl, J.E. *Unveiling the Universe: And Introduction to Astronomy*, Springer, 1996.

Weinberg, Steven. *Lectures of Astrophysics*, Cambridge University Press, 2020.

www.ingramcontent.com/pod-product-compliance
Lightning Source LLC
Chambersburg PA
CBHW070914290526
45795CB00001B/320